LONDON MATHEMATICAL SOCIETY LECTURE NOTE SERIES

Editor: PROFESSOR G. C. SHEPHARD, University of East Anglia

This series publishes the records of lectures and seminars on advanced topics in mathematics held at universities throughout the world. For the most part, these are at postgraduate level either presenting new material or describing older material in a new way. Exceptionally, topics at the undergraduate level may be published if the treatment is sufficiently original.

Prospective authors should contact the editor in the first instance.

Already published in this series

1. General cohomology theory and K-theory, PETER HILTON.
2. Numerical ranges of operators on normed spaces and of elements of normed algebras, F. F. BONSALL and J. DUNCAN.
3. Convex polytopes and the upper bound conjecture, P. McMULLEN and G. C. SHEPHARD.
4. Algebraic topology: A student's guide, J. F. ADAMS.
5. Commutative algebra, J. T. KNIGHT.
6. Finite groups of automorphisms, NORMAN BIGGS.
7. Introduction to combinatory logic, J. R. HINDLEY, B. LERCHER and J. P. SELDIN.
8. Integration and harmonic analysis on compact groups, R. E. EDWARDS.
9. Elliptic functions and elliptic curves, PATRICK DU VAL.
10. Numerical ranges II, F. F. BONSALL and J. DUNCAN.
11. New developments in topology, G. SEGAL (ed.).
12. Proceedings of the Symposium in Complex Analysis Canterbury 1973, J. CLUNIE and W. K. HAYMAN (eds.).
13. Combinatorics, Proceedings of the British Combinatorial Conference 1973, T. P. McDONOUGH and V. C. MAVRON (eds.).
14. Analytic theory of abelian varieties, H. P. F. SWINNERTON-DYER.
15. Introduction to topological groups, P. J. HIGGINS.

LONDON MATHEMATICAL SOCIETY LECTURE NOTE SERIES

[illegible]

London Mathematical Society Lecture Note Series. 16

Topics in Finite Groups

Terence M. Gagen

CAMBRIDGE UNIVERSITY PRESS

CAMBRIDGE

LONDON NEW YORK MELBOURNE

CAMBRIDGE UNIVERSITY PRESS
Cambridge, New York, Melbourne, Madrid, Cape Town, Singapore, São Paulo

Cambridge University Press
The Edinburgh Building, Cambridge CB2 8RU, UK

Published in the United States of America by Cambridge University Press, New York

www.cambridge.org
Information on this title: www.cambridge.org/9780521210027

First published 1976
Re-issued in this digitally printed version 2008

A catalogue record for this publication is available from the British Library

Library of Congress Catalogue Card Number: 75-17116

ISBN 978-0-521-21002-7 paperback

Contents

		page
Introduction		vii
Notations		viii
Elementary results		1
1.	Baer's Theorem	3
2.	A theorem of Blackburn	5
3.	A theorem of Bender	7
4.	The Transitivity Theorem	10
5.	The Uniqueness Theorem	12
6.	The case $\lvert\pi(F(H))\rvert = 1$	18
7.	The proof of the Uniqueness Theorem 5.1	20
8.	The Burnside $p^a q^b$-Theorem, p, q odd	30
9.	Matsuyama's proof of the $p^a q^b$-Theorem, $p = 2$	31
10.	A generalization of the Fitting subgroup	34
11.	Groups with abelian Sylow 2-subgroups	38
12.	Preliminary lemmas	40
13.	Properties of A*-groups	47
14.	Proof of the Theorem A, Part I	53
15.	Proof of the Theorem A, Part II	67
Appendix: p-constraint and p-stability		80
References		85

Introduction

The following material is selected from a course of lectures given at the University of Florida in Gainesville, Florida during 1971/72. The reader is expected to have read both Gorensteins' Finite Groups and much of Huppert's Endliche Gruppen I. In particular he must be familiar with the concepts of p-constraint and p-stability in order to begin, although there is a short discussion of these concepts in an appendix here.

The topics covered are such that I feel rather diffident about publishing these notes at all. The title should perhaps be changed to something like 'Lectures on some results of Bender on finite groups'. No less than three of his major results are studied here and of course the classification of A*-groups depends on his 'strongly embedded subgroup' theorem - which is not studied here at all. I feel that the theorems and techniques of the papers 'On the uniqueness theorem' and 'On groups with abelian Sylow 2-subgroups' are too important for finite groups and much too original to remain, as at present, accessible only to a very few specialists. I think that I understand the motivation for the abbreviation of the published versions of these two results. However, though it is clear that a proof becomes considerably more readable when a two or three page induction can be replaced by the words 'By induction we have', these details must sometime be filled in. And unfortunately, I think Dr. Bender has sometimes disguised the deepest and most elegant arguments by this very brevity. I hope that these notes will serve to make more of the group theoretical public aware of these incredibly rich results.

I must thank here the audience at the University of Florida - Mark Hale, Karl Keppler, Ray Shepherd and Ernie Shult. The contribution of Ernie Shult in particular cannot be minimized. Without him, we would all have floundered very soon.

December, 1973

Terry Gagen
Sydney, Australia

Notations

The notation used here is more or less standard. The reader should refer to [12] or [15] when in doubt.

$\mathcal{SCN}(P)$	The set of all self centralizing normal subgroups of P.
$\mathcal{SCN}(p)$	The set of all self centralizing normal subgroups of a Sylow p-subgroup.
$\mathcal{U}_G(A, \pi)$	The set of all A-invariant π-subgroups of G where π is a set of primes.
$\mathcal{U}^*_G(A, \pi)$	The maximal elements of $\mathcal{U}_G(A, \pi)$.
$r(P)$	The number of generators of an elementary abelian subgroup of P of maximal order (amongst all elementary abelian subgroups of P).
A^B	$\langle A^b : b \in B\rangle$.
G_p	A Sylow p-subgroup of G.
$0_\pi(G)$	The maximal normal π-subgroup of G, π a set of primes.
$0_{\sigma,\pi}(G)$	$0_\pi(G \bmod 0_\sigma(G))$.
$0^\pi(G)$	The smallest normal subgroup of G such that $G/0^\pi(G)$ is a π-group.
$F(G)$	The Fitting subgroup of G.
$\Phi(G)$	The Frattini subgroup of G.

The following two results are absolutely basic.

1. The Three Subgroups Lemma

   If $A, B, C \subseteq G$, $N \trianglelefteq G$ and $[A, B, C] \subseteq N$, $[B, C, A] \subseteq N$, then $[C, A, B] \subseteq N$.

2. If P is a p-group of class at most 2, then for all $n \in Z$ and for all $x, y \in P$,

$$(xy)^n = x^n y^n [y, x]^{n(n-1)/2}.$$

Elementary results

Definition. A group of automorphisms of a group P stabilizes a chain $P = P_0 \supseteqq P_1 \supseteqq \ldots \supseteqq P_n = 1$ if $[A, P_i] \subseteqq P_{i+1}$, $i = 0, \ldots, n-1$. Here $[a, x] = x^{-a}x$ for $x \in P$, $a \in A$.

Theorem 0.1. If a group of automorphisms A of a π-group P stabilizes a chain $P \supseteqq P_1 \supseteqq \ldots \supseteqq P_n = 1$, then A is a π-group.

Proof. Suppose $a \in A$ is a π'-automorphism of P. Clearly, by induction we may assume that $[a, P_1] = 1$. Then if $x \in P$, $x^a = xy$ where $y \in P_1$, since $[a, P] \subseteqq P_1$.

It follows that $x^{a^2} = xy^2, \ldots, x^{a^{|a|}} = xy^{|a|} = x$.

Since y is a π-element while a is a π'-element, we have that $y = 1$ and $[a, P] = 1$. Thus $a = 1$ and A is a π-group. //

Corollary 0.2. If A is a π'-group of automorphisms of a π-group P such that $[P, A, A] = 1$, then $[P, A] = 1$ and so $A = 1$.

Proof. A stabilizes the chain $P \supseteqq [P, A] \supseteqq [P, A, A] = 1$. //

Lemma 0.3. Let A be a π'-group of automorphisms of a π-group P. Let Q be an A-invariant normal subgroup of P. Then $C_{P/Q}(A) = C_P(A)Q/Q$.

Proof. Clearly $C_P(A)Q/Q \subseteqq C_{P/Q}(A)$.

Suppose now that xQ is a coset of Q in P which is fixed by A. Let QA act as a group of permutations on xQ where A acts in the obvious way and Q acts by multiplication on the right. Then QA acts transitively on xQ since Q does. Let A_1 be the stabilizer of a point. Then $|A_1| = |QA|/|xQ| = |A|$.

By the Schur Zassenhaus-Feit-Thompson Theorem A_1 is conjugate to A. Thus there exists $y \in xQ$ such that $y \in C_P(A)$. //

Remark. Note that in every application of 0.3, 0.4 in these notes, it will be known a priori that at least one of A or P is solvable. Hence the Feit-Thompson Theorem will not be required in the applications of 0.3, 0.4 here.

Corollary 0.4. _Let_ P _be a_ π_-group,_ A _a_ π'_-group of automorphisms of_ P. _Then_ $P = [P, A]C_P(A)$.

Proof. $[P, A] \subseteqq P$ and is A-invariant. Also A centralizes $P/[P, A]$. //

Corollary 0.5. _If_ A _is a_ π'_-group of automorphisms of an abelian_ π_-group_ P, _then_ $P = C_P(A) \oplus [P, A]$.

Proof. We show that $C_P(A) \cap [P, A] = 0$, writing P additively. Let θ be the endomorphism of P defined by $\theta = \frac{1}{|A|} \sum_{a \in A} a$. Clearly $b\theta = \theta b = \theta$ for all $b \in A$. Thus $\theta^2 = \theta$.

Since $P\theta \cap \ker \theta = 0$, $P = P\theta \oplus \ker \theta$.

Now if $x \in C_P(A)$, then $x\theta = \frac{1}{|A|} \sum_{a \in A} xa = x$ and so $C_P(A) \subseteqq P\theta$. Finally if $[x, a] \in [P, A]$, then $(-x + xa)\theta = -x\theta + x\theta = 0$ and so $[x, a] \in \ker \theta$. Thus $C_P(A) \cap [P, A] = 0$. The result follows from 0.4. //

Lemma 0.6. (Thompson) _Any p-group_ P _contains a characteristic subgroup_ C _such that_

(a) $cl\ C \leq 2$ _and_ $C/Z(C)$ _is elementary;_

(b) $[P, C] \subseteqq Z(C)$;

(c) $C_P(C) = Z(C)$;

(d) _Any automorphism_ $a \neq 1$ _of order prime to_ p _acts non-trivially on_ C.

Proof. We first show that (c) and C char P together ensure (d). For suppose that $[a, C] = 1$, where a is our given p'-automorphism. Then

$[a, C, P] = 1.$

Also

$$[C, P, a] \subseteq [C, a] = 1.$$

Thus

$$[P, a, C] = 1$$
$$[P, a] \subseteq C_P(C) = Z(C) \text{ by (c).}$$

Hence

$$[P, a, a] = 1. \text{ By 0.2, } [P, a] = 1.$$

To show the existence of C we proceed as follows. First if any subgroup $A \in \mathcal{SCN}(P)$ is characteristic in P, then take $A = C$. Clearly (a), (b), (c) hold. Hence we may suppose that no maximal abelian subgroup of P is characteristic. Let D be a maximal characteristic abelian subgroup of P. Clearly $C_P(D) \supset D$ and $C_P(D)$ char P. Let

$$C/D = \Omega_1(Z(P/D)) \cap C_P(D)/D.$$

Clearly $C \supset D$, and C is a characteristic subgroup of P. Since $D \subseteq Z(C)$ and $Z(C)$ is an abelian characteristic subgroup of P, maximality of D ensures that $D = Z(C)$. Clearly C is a characteristic subgroup of P.

(a) Since C/D is elementary abelian, first $C/Z(C)$ is elementary and then $[C, C] \subseteq D \subseteq Z(C)$. Hence $cl\ C \leq 2$.

(b) Since $C/D \subseteq Z(P/D)$, $[P, C] \subseteq D = Z(C)$.

(c) Suppose that $Q = C_P(C) \not\subseteq C$. Since $Q \cap C = Z(C) = D$ we have $Q/D \subset P/D$ and $Q/D \cap C/D = 1$. Of course $Q \subseteq C_P(C) \subseteq C_P(D)$. If $Q \neq D$, then Q/D intersects $\Omega_1(Z(P/D)) \cap C_P(D)/D$ non-trivially.

This contradiction completes the proof. //

1. BAER'S THEOREM

The Theorem 1.1 is required in the study of p-stable groups and a proof due to Suzuki is given in [12]. Of course, it follows immediately from the result of Baer [15] p. 298, this proof being given below as the first proof. Two other proofs of this result are given, both of which are interesting and brief.

Theorem 1.1 (R. Baer). Let K be a conjugacy class of p-elements in a finite group G. If $\langle x, y\rangle$ is a p-group for all $x, y \in K$, then $K \subseteqq O_p(G)$.

First Proof. Since $\langle x, y\rangle$ is a p-group for all $x, y \in K$, for all $g \in G$, $[x, g] = x^{-1}x^g$ and x are elements of the finite p-group $\langle x, x^g\rangle$. Hence

$$[g, x, x, \ldots, x] = 1 \text{ after a while.}$$

Thus x is a right Engel element and by Theorem III, 6.15 [15], $x \in F(G)$. Hence $K \subseteqq O_p(G)$. //

Second Proof (J. H. Walters). Let G be a minimal counter example to the Theorem. Let $M_1, M_2, \ldots, M_t$ be all the maximal subgroups of G containing a fixed element $x \in K$.

Clearly $O_p(G) = 1$ since G is a minimal counter example.

If $t = 1$, then for all $y \in K$, $\langle x, y\rangle \subseteqq M_1$, since $\langle x, y\rangle$ is a p-subgroup of G and so is certainly a proper subgroup of G containing x. Thus $K \subseteqq M_1$. Let $L = \langle K\rangle$. Then $L \subseteqq M_1 \subset G$ and $K \subseteqq O_p(L)$ by induction. Since $L \trianglelefteqq G$, we have a contradiction.

Thus we have $t > 1$. Among all i, j with $i \neq j$ choose $D = M_i \cap M_j$ such that $|D|_p$ is maximal. Let P be a Sylow p-subgroup of D containing x.

We show that there is no loss of generality in assuming that P is a Sylow p-subgroup of both M_i and M_j. For suppose that $P \subset P_i$, a Sylow p-subgroup of M_i. Then $N_G(P) \cap P_i \supset P$. Let M_k be a maximal subgroup of G containing $N_G(P) \subset G$. Then $M_k \cap M_i \supseteqq N_G(P) \cap P_i \supset P$ and so $k = i$ by the choice of i, j. Also $N_G(P) \subseteqq M_i$ and so P is a Sylow p-subgroup of M_j. Now choose $n \in N_G(P) \cap P_i - P$. Then clearly $n \notin M_j$ and so $M_j^n \neq M_j$. Otherwise $M_j \trianglelefteqq G$ and by induction $x \in K \cap M_j \subseteqq O_p(M_j) \subseteqq O_p(G) = 1$, a contradiction. Now $M_j^n \supseteqq P^n = P$ contains x and so $M_j^n = M_l$ for some l. Take $M_j \cap M_l$ as our required intersection. Note that P is a Sylow p-subgroup of both M_j and M_l.

We derive a contradiction easily now. By induction

$K \cap M_i \subseteq O_p(M_i) \subseteq P \subseteq M_j$.

Hence $K \cap M_i \subseteq K \cap M_j$.

Similarly $K \cap M_j \subseteq K \cap M_i$.

Thus $M_j = N_G(\langle K \cap M_j \rangle) = N_G(\langle K \cap M_i \rangle) = M_i$, a final contradiction. //

Third Proof (J. Alperin and R. Lyons). [1] Again let G be a minimal counter example. Let P be a Sylow p-subgroup of G. If $\langle K \rangle$ is a p-subgroup, then $K \subseteq O_p(G)$ since $K \trianglelefteq G$. Thus $\langle K \rangle$ is not a p-subgroup and so $K \nsubseteq P$. Let $y \in K - P$ and let Q be a Sylow p-subgroup of G containing y. Then of course $K \cap P \neq K \cap Q$.

Among all Sylow p-subgroups P, Q of G such that $K \cap P \neq K \cap Q$ choose P, Q so that $|K \cap P \cap Q|$ is maximal. Since $P^x = Q$ for some $x \in G$, $(K \cap P)^x = K \cap Q$ and so $K \cap P \nsubseteq Q$, $K \cap Q \nsubseteq P$. Let $D = \langle K \cap P \cap Q \rangle$. Suppose $D = P_0 \subset P_1 \subset \ldots \subset P_n = P$ where $[P_{i+1} : P_i] = p$.

Clearly $K \cap P \nsubseteq D$.

Suppose i is the smallest positive integer such that $K \cap P_i \nsubseteq K \cap D$. Let $x \in (K \cap P_i) - D$. Since $P_{i-1} \triangleleft P_i$, x normalizes P_{i-1} and so x normalizes $\langle K \cap P_{i-1} \rangle = D$. Choose $y \in (K \cap Q) - P$ similarly such that y normalizes D.

Then $\langle x, y \rangle$ is a p-group by hypothesis and so $\langle x, y, D \rangle$ is a p-group also. Let R be a Sylow p-subgroup of G containing $\langle x, y \rangle D$.

Then $\langle x, D \rangle \subseteq R \cap P$ implies that $R = P$ while $\langle y, D \rangle \subseteq R \cap Q$ implies that $R = Q$. This is a contradiction. //

2. A THEOREM OF BLACKBURN

This theorem duplicates some of the results of [12] - but its proof is so beautiful that it should be included here. The following lemma is of crucial importance for many of the results to come.

Lemma 2.1 (J. Thompson). Let a be a p'-automorphism of a p-group G. Suppose that X is a p-group of automorphisms of G and $[a, X] = [a, C_G(X)] = 1$. Then $a = 1$.

Proof. Let $N \subseteq G$ be X-invariant such that $[a, N] \neq 1$, but $[a, K] = 1$ for all X-invariant proper subgroups K of N. Then apply the

Three Subgroups Lemma, We have

$$[N, X, a] = 1 \text{ because } [N, X] \subset N$$

and is X invariant.

$$[X, a, N] = 1.$$

Thus $[a, N, X] = 1$, $[N, a] \subseteqq C_G(X)$, $[N, a, a] = 1$. By 0.2, $[N, a] = 1$. This completes the proof. //

Lemma 2.2. Let a be a π'-automorphism of a π-group G and suppose $X \lhd \lhd G$ is such that $[a, X] = [a, C_G(X)] = 1$. Then $a = 1$.

Proof. Let $X \lhd X_1 \lhd \ldots \lhd X_n = G$ and choose i such that $[a, X_{i+1}] \neq 1$, $[a, X_i] = 1$. Let $N = N_G(X_i)$. Since $X_{i+1} \subseteqq N$, $[a, N] \neq 1$. But

$$[X_i, N, a] = 1$$
$$[X_i, a, N] = 1.$$

Hence $[N, a, X_i] = 1$, $[N, a] \subset C_G(X_i) \subseteqq C_G(X)$. Thus $[N, a, a] = 1$. Lemma 0.2 implies that $[N, a] = 1$. //

Lemma 2.3 (N. Blackburn) [6]. Let a be a p'-automorphism of a p-group P. Let E be an abelian subgroup of P, maximal of exponent p^n, where $n \geqq 2$ if P is a non-abelian 2-group and no restriction is placed on n otherwise. If $[a, E] = 1$, then $a = 1$.

Proof. Let P be a minimal counter example.

If $C = C_P(E) \subset P$, then $[a, C] = 1$ by induction. By 2.2, $a = 1$. Thus $E \subseteqq Z(P)$.

Also, since $E \neq P$ trivially, $\Phi(P)E \subset P$. Thus a centralizes $\Phi(P)$ by induction. If $C(\Phi(P)) \subset P$, again we have $[a, \Phi(P)] = [a, C_P(\Phi(P))] = 1$ since $E \subset C(\Phi(P))$. By 2.2, $a = 1$ again. Thus P has class at most 2 and $\Phi(P) \subseteqq Z(P)$.

Choose $x \in P$ and consider $[x, a]^{p^n}$.

$$[x, a]^{p^n} = (x^{-1}x^a)^{p^n} = x^{-p^n}(x^a)^{p^n}[x^a, x^{-1}]^{p^n(p^n-1)/2}.$$

If P is abelian then of course $[x^a, x^{-1}] = 1$. On the other hand, if P is non-abelian, then

$$[x^a, x^{-1}]^{p^n(p^n-1)/2} = [x^a, x^{-p}]^{p^{n-1}(p^n-1)/2}$$

since if $p = 2$, $n \geqq 2$. But $x^{-p} \in \Phi(P) \subseteqq Z(P)$ for all $x \in P$. Thus in every case we have

$$[x, a]^{p^n} = x^{-p^n}(x^a)^{p^n} = x^{-p^n}(x^{p^n})^a.$$

Since $[a, \Phi(P)] = 1$ and $x^{p^n} \in \Phi(P)$, we see that $[x, a]^{p^n} = 1$. By the maximality of E, $[x, a] \in E$, for all $x \in P$. Thus $[P, a] \subseteqq E$ and $[P, a, a] = 1$. By 0.2, $[P, a] = 1$ and so $a = 1$. //

Theorem 2.4. <u>Let</u> P <u>be a p-group.</u> <u>If</u> a <u>is a p'-automorphism of</u> P <u>which centralizes</u> $\Omega_1(P)$ <u>then</u> $a = 1$ <u>unless</u> P <u>is a non-abelian 2-group.</u> <u>If</u> $[a, \Omega_2(P)] = 1$, <u>then</u> $a = 1$ <u>without restriction.</u> //

3. A THEOREM OF BENDER

Theorem 3.1 [2]. <u>Let</u> G <u>be a p-constrained group.</u> <u>If</u> $p = 2$ <u>assume that the Sylow 2-subgroups of</u> G <u>have class</u> $\leqq 2$. <u>Let</u> E <u>be an abelian p-subgroup of</u> G <u>which contains every p-element of its centralizer.</u> <u>Then every E-invariant p'-subgroup</u> H <u>of</u> G <u>lies in</u> $O_{p'}(G)$.

Remark 1. If E is a self centralizing normal subgroup of a Sylow p-subgroup of G, then E contains every p-element of its centralizer in G. For let $E \in \mathcal{SCN}(P)$ where P is a Sylow p-subgroup of G. Suppose that $D \supseteqq E$ is a Sylow p-subgroup of $C_G(E)$. Consider $N_G(E)$. Suppose that Q is a Sylow p-subgroup of $N_G(E)$ containing D. Since $P \subseteqq N_G(E)$, there exists $n \in N_G(E)$ such that $Q^n = P$. Then $D^n \subseteqq P \cap C_G(E) = E$. Hence $D = E$ is a Sylow p-subgroup of $C_G(E)$. By Burnside's Theorem, $C_G(E) = E \times O_{p'}(C_G(E))$.

Remark 2. Theorem 3.1 cannot hold without restriction if $p = 2$, even if G is solvable. Consider for example $G = GL(2, 3)$, $E \subseteqq G$ a fours-group. Then $O_{2'}(G) = 1$, $C_G(E) = E$, but there is a subgroup of order 3 which is normalized by E. Note that a Sylow 2-subgroup of GL(2, 3) has class 3.

Proof. Let G be a minimal counter example. The proof proceeds by a series of steps.

1. $O_{p'}(G) = 1$.

Otherwise, let $\overline{G} = G/O_{p'}(G)$. Since $C_{\overline{G}}(\overline{E}) = \overline{C_G(E)}$, by 0.3, we have $\overline{H} \subseteqq O_{p'}(\overline{G}) = 1$. Thus $H \subseteqq O_{p'}(G)$.

Let $R = O_p(G)$ and let $Q \neq 1$ be a minimal E-invariant p'-subgroup of G. If $RQE \subset G$, then $Q \subseteqq O_{p'}(RQE)$ by induction. Hence $[R, Q] \subseteqq O_{p'}(RQE) \cap R = 1$. Since $C_G(R) \subseteqq R$ by p-constraint we have

2. $G = RQE$.

Let S be a QE-invariant subgroup of G minimal with respect to $[Q, S] \neq 1$. Then S is a special p-group. The argument which verifies this is standard. See for example [12].

If S is abelian, then $S = C_S(Q) \oplus [Q, S]$ by 0.5. But $[Q, S]$ is an E-invariant p-subgroup and so $C_P(E) \cap [Q, S] \neq 1$. Thus $E \cap [Q, S] \neq 1$ by our hypothesis on E. On the other hand $[E \cap [Q, S], Q] \subseteqq Q \cap S = 1$. This contradicts $C_S(Q) \cap [Q, S] = 1$.

If S is non-abelian and p is odd, we use a remarkable idea of Bender, or perhaps of Baer. First by 2.4, since $[Q, S] \neq 1$, and S is minimal, $S = \Omega_1(S)$ has exponent p. Let T be a new group defined as follows: T = S qua set.

Every element $x \in S$ has a unique square root $x^{\frac{1}{2}} \in S$, since p is odd. Define a binary operation o on T as follows $x \circ y = x^{\frac{1}{2}} y x^{\frac{1}{2}}$.

It is routine to check that T is an elementary abelian group. Also QE acts as a group of automorphisms of T. Since S = T as sets, the fixed points of both Q and E on T are unchanged. But we have already reached a contradiction when S is abelian. This same argument can be applied to TQE.

If $p = 2$ and S is non-abelian, first $[E, S] \subseteqq Z(S)$ since SE has class $\leqq 2$. Thus

$$[S, E, Q] \subseteqq [Z(S), Q] = 1$$
$$[Q, S, E] \subseteqq [S, E] \subseteqq Z(S).$$

Hence

$$[E, Q, S] \subseteq Z(S).$$

If $[E, Q] \neq 1$, then $[E, Q] \subseteq Q$ stabilizes the chain $S \supseteq Z(S) \supseteq 1$. Hence $[E, Q] \subseteq C_G(S)$ by 0.1. Since $[E, Q]$ is an E-invariant subgroup of Q, minimality of Q ensures that $Q = [E, Q] \subseteq C_G(S)$. This is a contradiction.

Thus we may assume that $[E, Q] = 1$. Since $[S, E] \subset S$ is then Q-invariant, minimality of S ensures that $[S, E, Q] = 1$. Also $[E, Q, S] = 1$. It follows that $[Q, S, E] = [S, E] = 1$. Our assumptions on E now give $S \subseteq E$ and $[S, Q] \subseteq Q \cap S = 1$. This contradiction completes the proof. //

Lemma 3.2. Suppose P is a p-subgroup of a p-constrained group G. Then $O_{p'}(N_G(P)) \subseteq O_{p'}(G)$.

Proof. Since G is p-constrained, it follows from 0.3 that $G/O_{p'}(G)$ is p-constrained. Let $\overline{G} = G/O_{p'}(G)$ etc. By induction we have $O_{p'}(N_{\overline{G}}(\overline{P})) \subseteq O_{p'}(\overline{G}) = 1$. But clearly $O_{p'}(N_{\overline{G}}(\overline{P})) = O_{p'}(C_{\overline{G}}(\overline{P})) = \overline{O_{p'}(C_G(P))}$ by 0.3. Thus $O_{p'}(N_{\overline{G}}(\overline{P})) = \overline{O_{p'}(C_G(P))}$. Since $O_{p'}(N_G(P)) \subseteq C_G(P)$, it follows that $O_{p'}(N_G(P)) \subseteq O_{p'}(G)$ in this case. Hence we may assume that $O_{p'}(G) = 1$. Let $M = O_p(G)$, $Q = O_{p'}(N_G(P))$. Since $[Q, P] = 1$ and $[Q, C_M(P)] \subset M \cap Q = 1$ we have $[Q, M] = 1$ by 2.1. Hence $Q = 1$ because $C_G(M) \subseteq M$ by p-constraint.

Remark. The reader should refer to Lemma 12.5,12.6 due to Bender for a far reaching generalization of this result.

Lemma 3.3. If G is a p-solvable group of odd order and P is a Sylow p-subgroup of G such that $r(P) \leq 2$, then G has p-length 1.

Proof. Let G be a minimal counter example. Clearly $O_{p'}(G) = 1$. Let $R = O_p(G)$. Since G is p-constrained, $C_G(R) \subseteq R$. Let C be a Thompson critical subgroup of R and let $D = \Omega_1(C)$. Since $|G|$ is odd and C has a class ≤ 2, D has exponent p.

Since $r(P) \leq 2$, $r(D) \leq 2$. If $|Z(D)| \geq p^2$, then $D = Z(D)$ and if

$|Z(D)| = p$ then any subgroup of type (p, p) containing $Z(D)$ has centralizer of index $\leqq p$. It follows that $|D| \leqq p^3$. Also $|D| = p^3$ only if D is non-abelian of exponent p. Let $\overline{D} = D/\Phi(D)$. Then $C_G(\overline{D})$ is still a normal p-subgroup of G and so $C_G(\overline{D}) \subseteqq R$. But $G/C_G(\overline{D}) \subseteqq GL(2, p)$. But any odd order subgroup of $GL(2, p)$ has a normal Sylow p-subgroup.

Thus $G = O_{p,p'}(G \bmod C_G(\overline{D}))$ and so $G = O_{p,p'}(G)$. //

Lemma 3.4. <u>If G is a solvable group of odd order and P is a Sylow p-subgroup of G such that P' is cyclic, then G has p-length 1.</u>

Proof. Let G be a minimal counter example. First $O_{p'}(G) = 1$ clearly. Let $R = O_p(G)$. Then $C_G(R) \subseteqq R$ since G is solvable.

If $\Phi(R) \neq 1$, let $\overline{G} = G/\Phi(R)$. Since $\overline{P}'$ is cyclic, $\overline{G}$ has p-length 1. Let $Q\Phi(R) = O_{p'}(G \bmod \Phi(R))$ where Q is a p'-group. Then $[Q, R] \subseteqq \Phi(R)$ and so Q centralizes R modulo $\Phi(R)$. Thus $[Q, R] = 1$. Hence $O_{p'}(G \bmod \Phi(R)) = 1$ and G has p-length 1.

Thus we have that R is an elementary abelian p-group. Let $x \in P$. Then $[x, R] \subseteqq P' \cap R$, a cyclic subgroup of order p. Thus

$$[x, R] \subseteqq Z(P) \text{ and } [R, x, x] = 1.$$

Hence x acts on R with quadratic minimum polynomial. For a discussion of this see the Appendix p. 80. By the famous Theorem B of Hall and Higman $[x, R] = 1$. Thus $P = R$ and the Lemma is proved. //

4. THE TRANSITIVITY THEOREM

Included here is a proof of a rather unsatisfactory form of the Thompson Transitivity Theorem. This is proved completely in [12]. The proof given here is shorter but the Theorem is less general. More precisely, for the case of odd primes, the Theorem is more general; but for $p = 2$ it deals only with groups, whose Sylow 2-subgroups are of class at most 2. I do not know of any slick way to prove the general result. The final difficulty arises from the fact that a subgroup can be self centralizing and normal in one Sylow p-subgroup of a group but contained in another Sylow p-subgroup non-normally. The Theorem 3.1 is used to overcome this difficulty by replacing the elements of $\mathcal{SCN}$ by a rather larger class of groups. By this means we lose the case $p = 2$, class $\geqq 3$. No matter: the result as stated suffices for the results in these notes.

Before stating the Main Theorem, we prove a couple of auxiliary results.

Lemma 4.1. *If* P *is a p-subgroup of* G *such that* $N_G(P)$ *is p-constrained, then* $C_G(P)$ *is also p-constrained.*

Proof. First $O_{p'}(N_G(P)) = O_{p'}(C_G(P))$, clearly. Using 0.3, we may assume that $O_{p'}(N_G(P)) = 1$. Let $N = N_G(P)$, $C = C_G(P)$, $Q = O_p(C)$, $R = O_p(N)$. Since Q char $C \trianglelefteqq N$, $Q \subseteqq R \cap C$. Since $R \cap C \trianglelefteqq C$, $R \cap C \subseteqq Q$ and so $Q = R \cap C$.

Let $x \in C_C(Q)$ be a p'-element. Then $[x, R] \subseteqq R \cap C = Q$. Thus x stabilizes the chain $R \supseteqq Q \supseteqq 1$ and so by 0.1, $x \in C_N(R) \subseteqq R$ since N is p-constrained. It follows that $C_C(Q)$ is a p-subgroup, normal in C. Thus $C_C(R) \subseteqq Q$ and C is p-constrained. //

Lemma 4.2. *Suppose that* A *is an elementary abelian p-group such that* $r(A) = 3$. *If* P, Q *are A-invariant p'-groups, there exists* $a \in A$ *such that* $C_P(a) \neq 1$, $C_Q(a) \neq 1$.

Proof. Let V be a subgroup of A of type (p, p). Since $P = \langle C_P(v) : v \in V^{\#} \rangle$, there exists $v \in V^{\#}$ such that $C_P(v) \neq 1$. Let $W \subseteqq A$ such that W is of type (p, p) and $W \cap \langle v \rangle = 1$. There exists $w \in W$ such that $C_P(w) \cap C_P(v) \neq 1$, since $C_P(v)$ is W-invariant. Then $\langle v, w \rangle$ is of type (p, p) and acts on Q. Thus there exists an element $a \in \langle v, w \rangle^{\#}$ such that $C_Q(a) \neq 1$. Since $C_P(a) \supseteqq C_P(w) \cap C_P(v) \neq 1$, we are done. //

Theorem 4.3. (Transitivity Theorem) [8]. *Let* G *be a group in which the normalizer of every non-trivial p-subgroup is p-constrainted. If* $p = 2$, *assume that a Sylow p-subgroup of* G *has class* $\leqq 2$. *Let* E *be an abelian p-subgroup of* G *such that* $r(E) \geqq 3$ *and such that* E *contains every p-element of* $C = C_G(E)$. *Then* $O_{p'}(C)$ *acts transitively on* the elements of $\mathcal{U}^*(E, q)$ where q is a prime, $q \neq p$.

Proof. Let $\mathcal{S}_1, \mathcal{S}_2, \ldots, \mathcal{S}_t$ be the $O_{p'}(C)$ orbits of elements of $\mathcal{U}^*_G(E, q)$ and suppose that $t > 1$. Then clearly $\mathcal{U}^*_G(E, q) \neq \{1\}$.

Consider now $R = S_i \cap S_j$ for subgroups $S_i \in \mathcal{S}_i$, $S_j \in \mathcal{S}_j$, where $i \neq j$ and suppose that R is chosen of maximal order. For convenience, write $i = 1$, $j = 2$. Of course, $R \subset S_1$, $R \subset S_2$.

Now $N = N_G(R) \supseteqq E$. Consider $\overline{N} = N(R)/R$. Let $T_i = N \cap S_i \supset R$. Then $\overline{T}_i = T_i/R$, $i = 1, 2$ are E-invariant, non-trivial, and so by Lemma 4.2, there exists $e \in E^{\#}$ such that $C_{\overline{T}_i}(e) \neq 1$, $i = 1, 2$. By 0.3, $C_{\overline{T}_i}(e) = \overline{C_{T_i}(e)}$ and so $(C_G(e) \cap T_i)R \supset R$. By hypothesis and Lemma 4.1, $C_G(e) = H$ is p-constrained, and $C_G(e) \supseteqq E$.

Let $P_i = T_i \cap H$. Then $P_iR \supset R$. Remember $P_i \subseteqq N_G(R)$. Now P_i is E-invariant and by 3.1, $P_i \subseteqq O_{p'}(H)$.

Let $L = R(N \cap H)$. Since $P_i \subseteqq O_{p'}(H) \cap N \subseteqq O_{p'}(N \cap H)$, $P_i \subseteqq O_{p'}(L)$ because $R \trianglelefteqq L$ and R is a p'-group. Thus $RP_i \subseteqq O_{p'}(L)$, $i = 1, 2$.

Let $Q_i \supseteqq P_iR$ be an E-invariant Sylow q-subgroup of $O_{p'}(L)$, $i = 1, 2$. There exists $x \in C_G(E) \cap O_{p'}(L)$ such that $Q_1^x = Q_2$ by [12], 6.2.2. Since $\langle x \rangle$ is an E-invariant p'-subgroup of $N_G(E)$, which is p-constrained, 3.1 ensures that $x \in O_{p'}(N_G(E))$, $x \in O_{p'}(C_G(E))$.

Let $U \in \mathcal{U}^*_G(E, q)$, $U \supseteqq Q_1$. Then $U \cap S_1 \supseteqq Q_1 \cap S_1 \supseteqq P_1R \supset R$. By choice of R, $U \in \mathcal{S}_1$.

But $U^x \cap S_2 \supseteqq Q_1^x \cap S_2 = Q_2 \cap S_2 \supseteqq P_2R \supset R$. Now since $x \in C(E)$, U^x is a maximal element of $\mathcal{U}(E, q)$ clearly. By the choice of R, $U^x \in S_2$ and so $U \in \mathcal{S}_2$. Thus $\mathcal{S}_1 = \mathcal{S}_2$.

5. THE UNIQUENESS THEOREM

This and the next two sections are devoted to a proof, due to Bender, of the Uniqueness Theorem 5.1 [3].

Theorem 5.1. *Let G be a minimal simple group of odd order. Let U be an elementary abelian p-subgroup of G of order p^3. Then there is one and only one maximal subgroup of G containing U.*

Remark. This major theorem shortens much of the group theoretic Chapter IV of the Odd Order paper [8] of Feit and Thompson. It should be pointed out that if a group G of odd order contains no elementary abelian p-subgroup of order p^3 for any prime p, then G is not simple. In fact, such a group has an ordered Sylow tower, see [8].

If there is one and only one maximal subgroup M containing a subgroup $V \subseteq G$, then we call M a uniqueness subgroup.

Theorem 5.2. _Let G be a finite group, H a maximal local subgroup of G, $F = F(H)$. Let $X = C_H(F)$, $\pi = \pi(F)$. Assume that $|\pi| \geq 2$. Choose M a subgroup of F satisfying $C_F(M) \subseteq M$ and let R be a solvable subgroup of G normalized by MX. Then for any prime $q \in \pi'$, $\mathcal{N}_R(M, q)$ has only one maximal element and this is a Sylow q-subgroup of $O_{\pi'}(R)$._

Proof. Since $O_{\pi'}(R) \subseteq O_{\pi'}(RM) \subseteq O_{\pi'}(R)$, we may replace R by RM and assume that $M \subseteq R$. The proof is divided into a series of steps.

(i) If Y is a π'-subgroup of R normalized by M, then $Y \cap H = 1$.

(The reader should consult Theorem 12.4(a) for an appropriate generalization of this step.)

First $Y \cap H$ centralizes M since $[Y \cap H, M] \subseteq F(H) \cap Y = 1$. But $C_F(M) \subseteq M$. Now consider $(Y \cap H)F$. Let $p \in \pi$. If $x \in (Y \cap H)F$ is a p-element centralizing $M_{p'}$, then of course $x \in F_p$ and $x \in C_F(M_p) \cap C_F(M_{p'}) = C_F(M) \subseteq M$. Thus M_p has the property that it contains every p-element of its centralizer in $(Y \cap H)F$. Since this group is clearly p-constrained, we may apply 3.1 to get $Y \cap H \subseteq O_{p'}((Y \cap H)F)$ for all $p \in \pi$. Thus $[Y \cap H, F] = 1$ and so $Y \cap H \subseteq C_H(F)$.

Now $K = (Y \cap H)^X$ is a subgroup of R and so is a solvable X-invariant group. Thus $Y \cap H \subseteq S(X) \cap H \subseteq S(H) \cap C_H(F) \subseteq F$ by the well known property of Fitting subgroups of solvable groups. Since Y is a π'-group and F is a π-group, $Y \cap H = 1$.

(ii) If $Q \in \mathcal{N}_R(M, q)$, $q \in \pi'$, then $Q \subseteq O_{\pi'}(R)$.

We show $Q \subseteq O_{p'}(R)$ for all $p \in \pi$. For such a prime p, we have $M = M_p \times L$ and since $|\pi| > 1$ and $M \supseteq C_F(M)$ we must have $L \neq 1$.

Since $M_p \supseteq Z(F(H)_p)$, $C_G(M_p) \subseteq C_G(Z(F(H)_p)) \subseteq H$. Now $L \subseteq O_{p'}(H) \cap C_G(M_p) \subseteq O_{p'}(C_G(M_p)) \subseteq O_{p'}(N_G(M_p))$. By 3.2, $L \subseteq O_{p'}(R)$ since R is solvable and so $[L, Q] \subseteq O_{p'}(R)$. We show that $[L, Q] = Q$. For $Q = [L, Q]C_Q(L)$ by 0.4, and $C_G(L) \subseteq H$ since L contains a non-trivial normal subgroup of H, namely $Z(F(H)_{p'})$. Thus $C_Q(L) \subseteq Q \cap H = 1$ by step (i). Hence $Q = [Q, L] \subseteq O_{\pi'}(R)$.

We have thus shown in (i) and (ii) that every element of $\mathcal{V}_R(M, q)$, $q \in \pi'$, lies in $O_{\pi'}(R)$. On the other hand, some Sylow q-subgroup of $O_{\pi'}(R)$ is certainly M-invariant and the M-invariant Sylow q-subgroups of $O_{\pi'}(R)$ are conjugate under $C_G(M) \cap O_{\pi'}(R) = 1$. Thus $\mathcal{V}_R(M, q)$ has a unique maximal element. //

Theorem 5.3. Let G be a minimal simple group of odd order, H a maximal subgroup of G, U an elementary abelian p-subgroup of $F = F(H)$ such that

(i) $|U| \geq p^3$;

or (ii) $|U| = p^2$ and $U \subseteq A \in \mathcal{SCN}_3(p)$.

Set $M = C_F(U)$, $\pi(F(H))$. Assume $|\pi| \geq 2$. Then for any prime $q \in \pi'$, $\mathcal{V}_G(M, q) = 1$.

Proof. We show that $\mathcal{V}_G(M, q)$ has a unique maximal element. For then if Q is this unique maximal element of $\mathcal{V}_G(M, q)$, since $N_F(M)$ permutes the maximal elements of $\mathcal{V}_G(M, q)$ under conjugation, it follows that Q is the unique maximal element of $\mathcal{V}_G(N_F(M), q)$. After a while, since F is nilpotent, $\mathcal{V}_G(F, q)$ has a unique maximal element and then $\mathcal{V}_G(H, q)$ has a unique maximal element. But by the maximality of H, $Q \subseteq H$ and so $Q = 1$ because $O_q(H) = 1$.

The proof that $\mathcal{V}_G(M, q)$ has a unique maximal element follows closely the proof of the Transitivity Theorem 4.3.

Suppose that Q, R are maximal elements of $\mathcal{V}_G(M, q)$ and suppose Q, R have been chosen distinct such that $|Q \cap R|$ is maximal.

If $Q \cap R \neq 1$, let $K = N_G(Q \cap R)$. Apply 5.2 with K in place of R. We get $\mathcal{V}_K(M, q)$ has a unique maximal element S, say. Because $K \cap Q$, $K \cap R \in \mathcal{V}_K(M, q)$, $\langle K \cap Q, K \cap R \rangle \subseteq S$. Since $S \cap Q \supseteq K \cap Q \supset Q \cap R$, if $S^* \in \mathcal{V}_G^*(M, q)$, $S^* \supseteq S$, choice of Q, R ensures that $S^* = Q$. But again $S^* = R$, a contradiction. Thus $Q \cap R = 1$ and any two maximal elements of $\mathcal{V}_G(M, q)$ are disjoint.

Let $x \in U^\#$ and apply 5.2 to $C_G(x)$. It follows that $\mathcal{V}_{C_G(x)}(M, q)$ has a unique maximal element T. We show that there exists $x \in U$ such that $C_Q(x) \neq 1$, $C_R(x) \neq 1$. But $\langle C_Q(x), C_R(x)\rangle \subseteq T$. Let $T^* \in \mathcal{V}^*_G(M, q)$, $T^* \supseteq T$. Then $T^* \cap Q \supseteq C_Q(x) \neq 1$ whence $T^* = Q$. But again $T^* \cap R \supseteq C_R(x) \neq 1$ and so $T^* = R$. This is the required contradiction.

If $|U| \geq p^3$, we are clearly done by 4.2.

We may therefore assume that $|U| = p^2$ and $U \subseteq A \in \mathcal{SCN}_3(p)$. Clearly $\Omega_1(Z(F_p)) \subseteq U$ since otherwise we could replace U by $U\Omega_1(Z(F_p))$, an elementary abelian group of order $\geq p^3$. Thus $C_G(U) \subseteq C_G(\Omega_1(Z(F_p))) \subseteq H$.

Now $Q = \langle C_Q(x) : x \in U^\#\rangle$. As already noticed, $C_G(x)$ has a unique maximal M-invariant q-subgroup, X say. If $x \in U^\#$ is such that $C_Q(x) \neq 1$, then letting $X^* \in \mathcal{V}^*_G(M, q)$, $X^* \supseteq X$, we have $X^* \cap Q \supseteq C_Q(x) \neq 1$ and so $X^* = Q$. Thus $X = C_Q(x)$. Now A normalizes $M = C_F(U)$. Hence A permutes the elements of $\mathcal{V}_G(M, q)$ under conjugation. Thus $A \subseteq C_G(x)$ normalizes $C_Q(x)$ whenever $C_Q(x) \neq 1$. It follows that A normalizes Q. Similarly A normalizes R.

But by 4.3, Q and R are conjugate by an element of $O_{p'}(C_G(A))$. Since $C_G(A) \subseteq C_G(U) \subseteq H$, there exists $h \in C_H(A)$ such that $\langle Q^h, R\rangle$ is a q-group. But $h \in C_G(U)$ normalizes M. Hence M normalizes Q^h, R and so M normalizes $\langle Q^h, R\rangle \supseteq R$. It follows that $Q^h \subseteq R$ and so $Q^h = R$. Now if $u \in U$ is such that $C_Q(u) \neq 1$, then $C_R(u^h) \neq 1$. Since $h \in C_G(U)$, $u^h = u$. Hence there is an element $u \in U$ such that $C_Q(u) \neq 1$, $C_R(u) \neq 1$. This completes the proof. //

Theorem 5.4. <u>Let G be a minimal simple group of odd order, M a subgroup of F(H) containing Z(F(H)), $\pi = \pi(F(H))$. Assume that $|\pi| \geq 2$ and $\mathcal{V}_G(M, \pi') = \{1\}$. Then H is the only maximal subgroup of G which contains</u> M.

Proof. Suppose that L is a maximal subgroup of G which contains M. Clearly $\pi(F(L)) \subseteq \pi$ since $\mathcal{V}_G(M, \pi') = 1$. Because $Z(F(H)) \subseteq M$, the centralizer of any Hall subgroup of M is contained in

M, as we have seen before. Then if $\sigma \subseteq \pi$,

$M_{\sigma'} \subseteq C_G(M_p) \cap O_{\sigma'}(H) \subseteq O_{\sigma'}(C_G(M_p))$, where $p \in \sigma$.

Therefore $M_{\sigma'} \subseteq O_{p'}(C_G(M_p)) \cap L \subseteq O_{p'}(C_L(M_p))$ for all $p \in \sigma$.

By 3.2, $M_{\sigma'} \subseteq O_{p'}(L)$ for all $p \in \sigma$. Hence $M_{\sigma'} \subseteq O_{\sigma'}(L)$. Thus $M_{\sigma'} \subseteq C_G(F(L)_\sigma)$ and $F(L)_\sigma \subseteq C_G(M_{\sigma'}) \subseteq H$, for every subset $\sigma \subseteq \pi$. If $\sigma = \pi(F(L)) \neq \pi$, we see that $M_{\sigma'} \subseteq C_G(F(L)) \subseteq F(L)$, a σ-group. Thus $\pi = \pi(F(L))$.

Taking $\sigma = \pi - p$, we now have $M_{\sigma'} = M_p \subseteq O_p(L)$ centralizes $F(O_{p'}(L)) \supseteq C_{O_{p'}(L)}(F(O_{p'}(L)))$. Thus by 2.2, $[M_p, O_{p'}(L)] = 1$ and so $O_{p'}(L) \subseteq H$.

But $O_p(L) \subseteq F(L)_p \subseteq H$ and $C_G(O_p(L)) \subseteq L$. Hence $O_{p'}(L) \subseteq O_{p'}(C_G(O_p(L))) \cap H \subseteq O_{p'}(C_H(O_p(L)))$. By 3.2, $O_{p'}(L) \subseteq O_{p'}(H)$.

By symmetry, $O_{p'}(H) \subseteq O_{p'}(L)$ and since $O_{p'}(H) \neq 1$ we have $H = N_G(O_{p'}(H)) = N_G(O_{p'}(L)) = L$. //

Theorem 5.5. _Let_ G _be a minimal simple group of odd order,_ p _a prime,_ H _a maximal subgroup of_ G _satisfying_ $O_{p'}(H) \neq 1$, V _an elementary abelian subgroup of order_ p^2 _of_ G _such that_ $C_G(x) \subseteq H$ _for all_ $x \in V^{\#}$. _Then_ H _is the only maximal subgroup of_ G _containing_ V.

Proof. Let P be a Sylow p-subgroup of H containing V. Every p'-subgroup of G normalized by V is contained in H. Hence the subgroup $\langle \mathcal{N}_G(P, p')\rangle \subseteq H$. Let

$$P_1 = P \cap O_{p',p}(H), \quad Q \in \mathcal{N}_G(P, p'), \quad Q \subseteq H.$$

Then $[P_1, Q] \subseteq Q \cap O_{p',p}(H) \subseteq O_{p'}(H)$. Thus

$$QO_{p'}(H)/O_{p'}(H) \subseteq C_{H/O_{p'}(H)}(P_1) = C_H(P_1)O_{p'}(H)/O_{p'}(H).$$

Since H is a p-constrained, $C_H(P_1) \subseteq O_{p',p}(H)$. Thus $Q \subseteq O_{p',p}(H)$ and so $Q \subseteq O_{p'}(H)$. It follows that $\langle \mathcal{N}_G(P, p')\rangle \subseteq O_{p'}(H) \in \mathcal{N}_G(P, p')$. Hence $\langle \mathcal{N}_G(P, p')\rangle = O_{p'}(H)$.

Now $N_G(P)$ permutes the elements of $\mathcal{N}_H(P, p')$ under conjugation and so normalizes $\langle \mathcal{N}_G(P, p')\rangle = O_{p'}(H) \neq 1$.

Thus $N_G(P) \subseteq H$. Hence P is a Sylow p-subgroup of G.

Choose $L \supseteq V$, $L \neq H$ such that

(i) $|L \cap H|_p$ is maximal and then

(ii) $|L|_p$ is maximal and then

(iii) $|L|$ is maximal.

Let R be a Sylow p-subgroup of $L \cap H$ containing V. There is no loss of generality in assuming $R \subseteq P$ replacing V by an H-conjugate if necessary. If $R = P$, then $N_G(R) \subseteq H$. If $R \subset P$, then $N_P(R) \supset R$ and by the choice of L, H is the only maximal subgroup of G containing $N_P(R)$. Thus $N_G(R) \subseteq H$ in every case. Clearly R is a Sylow p-subgroup of L.

We have $O_{p'}(L) \subseteq H$ because $O_{p'}(L)$ is V-invariant.

Set $S = R \cap O_{p',p}(L) \trianglelefteq R$. Since $L = O_{p'}(L)\, N_L(S)$ and $O_{p'}(L) \subseteq H$, $L \neq H$, $N_G(S) \nsubseteq H$. Thus $N_P(S) = R$ and $S \neq R$. Because $|L|_p \leq |N_G(S)|_p$, we may assume that $N_G(S) \subseteq L$.

Now a solvable group of odd order with a Sylow p-subgroup of rank ≤ 2 has p-length 1 by 3.3. Since L does not have p-length 1, $r(R) \geq 3$. Thus $P \supseteq R$ contains a 3-generated abelian subgroup and by [8] Lemma 8.4, $\mathcal{SCN}_3(P) \neq \emptyset$. Let $A \in \mathcal{SCN}_3(P)$.

If $x \in N_A(S) \subseteq L$, then $[S, x, x] = 1$. The p-stability of L, a solvable group of odd order, shows that $x \in O_p(N_L(S) \bmod C_L(S)) = O_{p',p}(L)$. Thus $N_A(S) \subseteq S$ and $A \subseteq S$.

Now let q be a prime different from p. The Transitivity Theorem 4.3 ensures that if $Q \in \mathcal{U}^*_G(A, q)$, $n \in N_G(A)$, then $Q^n \in \mathcal{U}_G(A, q)$ and even $Q^n \in \mathcal{U}^*_G(A, q)$ because all elements of $\mathcal{U}^*_G(A, q)$ have the same order. Moreover, there exists $c \in O_{p'}(C_G(A))$ such that $Q^{nc} = Q$. Hence $N_G(A) = (N_G(Q) \cap N_G(A))\, O_{p'}(C_G(A))$. But $P \subseteq N_G(A)$ is a Sylow p-subgroup of $N_G(A)$ and so there exists $m^{-1} \in N_G(A)$ such that $P^{m^{-1}} \subseteq N_G(A) \cap N_G(Q)$.

Then Q^m is P-invariant. Thus if $Q \in \mathcal{U}^*_G(A, q)$, there exists a conjugate $Q^m \in \mathcal{U}^*_G(P, q)$. We have already seen that $\langle \mathcal{U}^*_G(P, q)\rangle \subseteq O_{p'}(H)$.

Thus at least one element of $\mathcal{U}^*_G(A, q)$ is contained in $O_{p'}(H)$.

But $C_G(A) = A \times O_{p'}(C_G(A))$ and $O_{p'}(C_G(A))$ char $C_G(A)$. Now V normalizes A, $C_G(A)$, $O_{p'}(C_G(A))$ and so $O_{p'}(C_G(A)) \subseteq H$. Since

$O_{p'}(C_G(A))$ acts transitively on the elements of $\mathcal{M}^*_G(A, q)$ and one of them lies in $O_{p'}(H)$, $\langle \mathcal{M}^*_G(A, q)\rangle \subseteqq O_{p'}(H)$. Since $S \trianglelefteq A$, $\mathcal{M}^*_G(S, p') = O'_p(H)$ and $N_G(S) \subseteqq N_G(\mathcal{M}^*_G(S, p')) = H$, a contradiction. //

6. THE CASE $|\pi(F(H))| = 1$

The Theorems 5.2, 5.3, 5.4, 5.5 are directed towards the case $|\pi| \geqq 2$. When $F(H)$ is a p-group, it was very soon recognized that similar results were implied by the Glauberman ZJ-Theorem.

Let G be a minimal simple group of odd order, H a maximal subgroup of G containing a Sylow p-subgroup P of H for which $\mathcal{SCN}_3(P) \neq \emptyset$. Let $A \in \mathcal{SCN}_3(P)$ and assume that $F(H)$ is a p-group. The first observation is that P is in fact a Sylow p-subgroup of G and moreover $N_G(P) \subseteqq H$.

Lemma 6.1. $\mathcal{M}_G(A, p') = \{1\}$ <u>and</u> $N_G(P) \subseteqq H$.

Proof. Since H is p-stable, any element $A \in \mathcal{SCN}_3(P)$ is contained in $O_p(H) = F(H) \supseteqq C_G(F(H))$. Thus $Z(F(H)) \subseteqq A$ and $C_G(A) \subseteqq C_G(Z(F(H))) \subseteqq H$. But by 3.2 $O_{p'}(C_H(A)) \subseteqq O_{p'}(H) = 1$ and so $C_G(A) = A$. By the ZJ-Theorem of Glauberman [9],

$$Z(J(P)) \trianglelefteq H. \quad \text{Hence } N_G(P) \subseteqq N(Z(J(P))) = H.$$

Now take any prime $q \neq p$. From the Transitivity Theorem 4.3, since $C_G(A) = A$, $\mathcal{M}_G(A, q)$ has only one maximal element. Since $A \trianglelefteq F(H)$, $\mathcal{M}^*_G(F(H), q)$ has only one element and then $\mathcal{M}^*_G(H, q)$ has only element Q. Since H is maximal in G, $Q \subseteqq H$ and then $Q \subseteqq O_p(H) = 1$. Thus $\mathcal{M}_G(A, q) = 1$ and so $\mathcal{M}_G(A, p') = 1$. //

Lemma 6.2. H <u>is a uniqueness subgroup for</u> P.

Proof. For if L is another maximal subgroup of G containing P, 6.1 implies that $O_{p'}(L) = 1$ and then the ZJ-Theorem ensures that $L = N(Z(J(P))) = H$. //

More surprising we see that H is in fact a uniqueness subgroup for A.

Lemma 6.3. H is a uniqueness subgroup for A.

Proof. Suppose L is another maximal subgroup of G containing A. Choose L in such a way that $|H \cap L|_p$ is maximal. Let Q be a Sylow p-subgroup of $H \cap L$ containing A. If Q is a Sylow p-subgroup of H, then $Q = P^h$, $h \in H$, and so $N_G(Q) \subseteqq H$. If $|Q| < |H|_p$, then $N_G(Q) \subseteqq H$ by the choice of L. Thus in any case Q is a Sylow p-subgroup of L.

Since $O_{p'}(L) = 1$ by 6.1, $L = N_G(Z(J(Q)))$. But then Q is a Sylow p-subgroup of G and also of H. This contradicts 6.2. //

Lemma 6.4. Let V be an elementary abelian subgroup of order p^2 of H and suppose that $C_G(x) \subseteqq H$ for all $x \in V^{\#}$. Then H is a uniqueness subgroup for V.

Proof. Suppose that L is another maximal subgroup of G containing V and choose L so that $|L \cap H|_p$ is maximal. Let Q be a Sylow p-subgroup of $H \cap L$ containing V. Again Q is a Sylow p-subgroup of L. By the ZJ-Theorem $L = O_{p'}(L)N_L(Z(J(Q)))$. Since by hypothesis $O_{p'}(L) \subseteqq H$, $N(Z(J(Q))) \nsubseteq H$. Since $|N_H(Z(J(Q)))|_p > |Q|_p$ we have a contradiction. //

Lemma 6.5. Let $X \subseteqq Y \subseteqq P$ where Y is an elementary abelian group of order p^3, $|X| = p^2$. Then H is a uniqueness subgroup for X.

Proof. Let V_1 be a Y-invariant subgroup of $\Omega_1(A)$ of order p^2. Let $V_2 = C_Y(V_1)$. Since $Y/C_Y(V_1)$ is a subgroup of GL(2, p), it has order at most p. Thus $|V_2| \geqq p^2$. If $x \in V_1^{\#}$, $C_G(x) \supseteqq A$, and since H is a uniqueness subgroup for A, $C_G(x) \subseteqq H$. By 6.4, H is a uniqueness subgroup for V_1. Now if $x \in V_2^{\#}$, $C_G(x) \supseteqq V_1$ and so $C_G(x) \subseteqq H$. By 6.4 again, H is a uniqueness subgroup for V_2. Finally, if $x \in X$, $C_G(x) \supseteqq Y \supseteqq V_2$ and so $C_G(x) \subseteqq H$. By 6.4 yet again, H is a uniqueness subgroup for X. //

7. THE PROOF OF THE UNIQUENESS THEOREM 5.1

We commence the final attack on 5.1 having established in §5 and §6 uniqueness theorems sufficient for the task. Thus we fix G a minimal simple group of odd order, P a Sylow p-subgroup of G for which $\mathcal{SCN}_3(P) \neq \emptyset$, H a maximal subgroup of G containing $N_G(P)$, if P is abelian, and containing $N_G(Z(P) \cap P') \supseteq N_G(P)$ otherwise. Let $A \in \mathcal{SCN}_3(P)$ and $A_0 = \Omega_1(A)$.

We show that H is a uniqueness subgroup for A and establish 5.1; but note that we will have in fact established that H is a uniqueness subgroup for every elementary abelian p-subgroup of order p^2 which lies in an elementary abelian p-subgroup of P of order p^3. In fact this is precisely the statement of 6.5 if $F(H)$ is a p-group, while if not, suppose that H is a uniqueness subgroup for $A \subseteq P$. Let $X \subseteq Y \subseteq P$ where Y is of type (p, p, p), $|X| = p^2$. Let Y normalize $V_1 \subseteq \Omega_1(A)$ of order p^2 and let $V_2 = C_Y(V_1)$. Then if $x \in V_1^{\#}$, $C_G(x) \supseteq A$ and so $C_G(x) \subseteq H$. By 5.5, H is a uniqueness subgroup for V_1. Now if $x \in V_2^{\#}$, $C_G(x) \supseteq V_1$ and so $C_G(x) \subseteq H$. We use here 5.5 since V_2 is non-cyclic to get that H is a uniqueness subgroup for V_1. Finally if $x \in X^{\#}$, $C_G(x) \supseteq V_1$ and so $C_G(x) \subseteq H$. By 5.5 again, H is a uniqueness subgroup for X.

We thus show that H is a uniqueness subgroup for A. We say that a maximal subgroup X of G is of <u>uniqueness type</u> and q is a <u>uniqueness prime</u> if there exists a prime q such that $F(X)$ has an elementary abelian q-subgroup U such that

(i) $|U| \geq q^3$ or

(ii) $|U| = q^2$ and $U \subseteq A \in \mathcal{SCN}_3(q)$.

Assume that H is not a uniqueness subgroup for A. Note that if $F(H)$ is an r-group where a Sylow r-subgroup R of H satisfies $\mathcal{SCN}_3(R) \neq \emptyset$, then H is a uniqueness subgroup for every elementary abelian subgroup U of R of order r^2 which lies in an abelian subgroup V or R of type (r, r, r) by 6.5.

Suppose on the other hand, $|\pi(F(H))| \geq 2$. Then, if U is an elementary abelian r-subgroup of $F(H)$ of order $\geq r^3$ or of order r^2 contained in an element of $\mathcal{SCN}_3(r)$, it follows from 5.3 that $\mathsf{M}_G(M, \pi') = \{1\}$ where $M = C_F(U)$, $F = F(H)$. Then by 5.4, H is a uniqueness subgroup for M. Now if $V \subseteq U$ is a subgroup of order r^2 and $x \in V^{\#}$, $C_G(x) \supseteq M$ from which it follows that $C_G(x) \subseteq H$. Now apply 5.5 to get that H is a uniqueness subgroup for V.

Thus a subgroup H of uniqueness type is a uniqueness subgroup for very many subgroups of H.

The proof of 5.1 hinges on the study of subgroups of uniqueness type. First we show that subgroups of uniqueness type exist in G.

Lemma 7.1. H is a subgroup of uniqueness type.

Proof. Suppose that H is not of uniqueness type. Then F(H) does not contain an elementary abelian subgroup of order q^3 for any prime q and a Sylow p-subgroup of F(H) is cyclic. For if $F(H)_p$ is non-cyclic, since $F(H)_p \cap P \trianglelefteq P$, there exists $V \trianglelefteq P$ of type (p, p), $V \subseteq F(H)_p$. But every such normal subgroup lies in an element of $\mathcal{SCN}_3(P)$. For let $A \in \mathcal{SCN}_3(P)$, $A_1 \subseteq A$ a normal subgroup of P of type (p, p, p). Since $C_H(V) \cap A_1$ has order $\geq p^2$, $V(C_H(V) \cap A_1) \subseteq P$ and is elementary abelian. If $C_H(V) \cap A_1 \nsubseteq V$, then V lies in a $\geq$ 3-generated abelian normal subgroup of P while if $C_H(V) \cap A_1 \subseteq V$, $V \subseteq A_1$.

Let $Q = F(H)_q$, where q is a prime $q \neq p$. If P does not centralize Q, let $R = \Omega_1(C)$ where C is a Thompson subgroup of Q. (See 0.4.) By 2.4, $[P, R] \neq 1$. Since $r(Q) \leq 2$, $r(R) \leq 2$ and so $|R| \leq q^3$. Let $\bar{R} = R/\Phi(R)$. Then P acts non-trivially on $\bar{R}$ and $H/C_H(\bar{R})$ is an odd order subgroup of GL(2, q). It follows that $(H/C_H(\bar{R}))'$ is a q-group and so $P \cap H' \subseteq C_H(\bar{R})$. Thus $P \cap H' \subseteq C_H(Q)$. This holds for all $q \neq p$. Since $H/C_H(F(H)_p)$ is abelian because $F(H)_p$ is cyclic, $P \cap H' \subseteq C_H(F(H)_p)$ and $P \cap H' \subseteq C_H(F(H)) \subseteq F(H)$.

If P is abelian, then fusion of elements of P is controlled by $N_G(P)$ by the well-known Burnside Lemma. Thus the focal subgroup

$$\langle x^{-1}x^g : x \in P,\ g \in G,\ x^g \in P\rangle = S \text{ satisfies}$$

$$S = \langle x^{-1}x^n : x \in P,\ n \in N_G(P)\ (\subseteq H)\rangle \subseteq P \cap H' \subseteq F(H).$$

Thus by the application of the transfer we have $P \cap G' \subseteq F(H)$.

Since G has no non-trivial p-factor group, $P \subseteqq F(H)$ is cyclic. This is not the case.

Hence $P' \neq 1$. But since $P \cap H' \subseteqq F(H)$ we must have $P \cap F(H) \neq 1$. Let Z be the subgroup of order p contained in $F(H)$. If $Z \subseteqq H \cap H^g$ where $g \in G - H$, choose g so that $|H \cap H^g|$ is maximal. Let S be a Sylow p-subgroup of $H \cap H^g$ and assume without loss of generality that $S \subseteqq P$. Since $Z \trianglelefteq H$, Z lies in every Sylow p-subgroup of H. If $S = P \subseteqq H \cap H^g$ then $P, P^{g^{-1}} \subseteqq H$ and so $P = P^{g^{-1}h}$, $h \in H$. But then $g^{-1}h \in N_G(P) \subseteqq H$, $g \in H$, a contradiction. Therefore $S \subset P$.

Let T be a Sylow p-subgroup of $N_G(S)$. Then $T \supset S$ and so $N_G(T) \subseteqq H$ by the choice of H^g. Hence T is a Sylow p-subgroup of $N_G(S)$. Since $P \cap H' \subseteqq F(H)$, $T' \subseteqq F(H)$ and so T' is cyclic. By 3.4, $N_G(S)$ has p-length 1.

Thus $N_G(S) = O_{p'}(N_G(S))(N_G(S) \cap N_G(T))$ by the Frattini argument. Since $N_G(S) \nsubseteqq H$, $N_G(T) \subseteqq H$, $O_{p'}(N_G(S)) \nsubseteqq H$. But $O_{p'}(N_G(S)) \subseteqq C_G(S) \subseteqq C_G(Z) \subseteqq H$. This contradiction shows that if $Z \subseteqq H \cap H^g$, then $g \in H$.

Now let $x \in P$, $x^g \in P$. Then $Z \subseteqq Z(P)$ and so $Z, Z^{g^{-1}} \subseteqq C_G(x)$. Choose $y \in C_G(z)$ such that $\langle Z, Z^{g^{-1}y}\rangle$ is a p-group and then find $z \in G$ such that $\langle Z, Z^{g^{-1}y}\rangle^z \subseteqq P$. Then we have first $Z, Z^z \subseteqq P$ whence $z \in H$ and then $Z, Z^{g^{-1}yz} \in P$ whence $g^{-1}yz \in H$. Thus $g^{-1}y \in H$. Hence $x^g = x^{y^{-1}g}$ where $y^{-1}g \in H$ since $y \in C_G(x)$. Thus H controls p-fusion.

Transfer now yields $P \cap G' \subseteqq P \cap H' \subseteqq F(H)$. This contradicts the simplicity of G. //

We devote the next several lemmas to a study of subgroups of uniqueness type, now being certain that they exist. Fix our notation so that X is a subgroup of uniqueness type, q is a uniqueness prime. Let $B = \Omega_1(Z_2(F(X)_q))$. Remember that H is not a uniqueness subgroup for A, otherwise we are done.

Lemma 7.2. Either X is a uniqueness subgroup for every subgroup of order q^2 of B or B is non-abelian of order q^3 and X is a uniqueness subgroup for every subgroup of order q^2 of B which is *normal in some Sylow* q-*subgroup of* X.

Proof. If $|B| \geqq q^4$ then every subgroup V of order q^2 of B lies in an elementary abelian subgroup U of order q^3 of B. For if $Z(B) \not\subseteq V$, the result is clear and if $Z(B) \subseteqq V$ either $V = Z(B)$ in which case the result is trivial using $\exp B = q$, or $Z(B) \subset V$, $|Z(B)| = q$. But then $V \trianglelefteq B$ since $B' = Z(B)$ and so $B/C_B(V)$ has order $\leqq q$. Thus $|C_B(V)| \geqq q^3$ and we are done.

If $\pi(F(X)) = \{q\}$, then 6.5 shows that H is a uniqueness subgroup for V. If $\pi(F(X)) \supset \{q\}$, then 5.4 implies that $\mathcal{N}_G(M, \pi') = 1$, where $M = C_{F(X)}(U)$. Then 5.5 implies that X is a uniqueness subgroup for M. Now if $x \in V^{\#} \subseteqq U$, $C_G(x) \supseteqq M$ and so $C_G(x) \subseteqq X$. Thus X is a uniqueness subgroup for V by 5.6.

We may therefore assume that $|B| \leqq q^3$. If B is abelian of order q^3 the above argument applies. We have already remarked that X is a uniqueness subgroup for some q-subgroups of X and so X must contain a Sylow q-subgroup Q of G! Now any normal subgroup V of Q of type (q, q) lies in an element of $\mathcal{SCN}_3(Q)$ by a familiar argument. Thus by the remark at the beginning of §7 we see that X is a uniqueness subgroup for V. This completes the proof. //

Lemma 7.3. *If* $V \subseteqq Y \subseteqq X$, *where* Y *is of type* (q, q, q) *and* $|V| = q^2$, *then* X *is an uniqueness subgroup for* V.

Proof. By 7.2, Y normalizes a q-group U of order q^2 and type (q, q) for which X is an uniqueness subgroup. Thus $|C_Y(U)| \geqq q^2$ since $Y/C_Y(U) \subseteqq GL(2, q)$. If $x \in C_Y(U)^{\#}$, then $C_G(x) \supseteqq U$ and so $C_G(x) \subseteqq X$. By 6.6 and 5.5, X is a uniqueness subgroup for $C_Y(U)$. Thus X is a uniqueness subgroup for V by 6.6 and 5.5 again, since if $x \in V^{\#}$, $C_G(x) \supseteqq Y \supseteqq C_Y(U)$ and so $C_G(x) \subseteqq X$. //

Lemma 7.4. *The uniqueness prime* q *belonging to* X *is different from* p.

Proof. This is clear by 7.3 since X is not a uniqueness subgroup for A. //

Lemma 7.5. *No non-cyclic subgroup of A centralizes a non-cyclic subgroup of B.*

Proof. Let $D \subseteqq A_0$ be of order p^2 and suppose that $V \subseteqq C_B(D)$ is of type (p, p).

If B is non-abelian of order q^3, $V/Z(B)$ is a non-trivial subgroup of $B/Z(B)$ which is centralized by D. But the elements of D induce automorphisms of B which preserve the symplectic form $[\,,\,] : B/Z(B) \times B/Z(B) \to Z(B)$ and so these automorphisms have determinant 1. Thus if B is non-abelian of order q^3, then $|C_B(D)| \geqq p^2$ implies that $C_B(D) = B$ and so X is a uniqueness subgroup for $C_B(D)$ by 7.2. If B is not non-abelian of order q^3, then X is a uniqueness subgroup for $C_B(D)$ by 6.2 also.

We may now apply Lemma 5.6. Note that by 7.4, $O_{p'}(X) = 1$, and if $x \in D^{\#}$, $C_G(x) \supseteqq C_B(D)$. Since X is a uniqueness subgroup for $C_B(D)$, $C_G(x) \subseteqq X$. By 5.6, X is a uniqueness subgroup for D, and also A, and this is a contradiction. //

Lemma 7.6. *Let $A_0 = \Omega_1(A)$ and $B = \Omega_1(Z_2(F(X)_q))$ as usual. Then*

(i) $C_B(A_0) = 1$ *and* $|A_0| = p^3$

(ii) A_0 *contains a subgroup* D *of order* p^2 *such that if* $E = C_B(D)$, *then* $|E| = q$ *and* $N_G(E) \subseteqq X$

(iii) *There exists* $d \in D^{\#}$ *such that* $C_G(d) \subseteqq X$.

Proof. By 7.5, if D is any subgroup of A_0 or order p^2, $|C_B(D)| \leqq q$.

If B is non-abelian of order q^3, then A_0 acts on $\overline{B} = B/Z(B)$ and so $A_0/C_{A_0}(\overline{B}) \subseteqq GL(2, q)$. Thus $|A_0/C_{A_0}(\overline{B})| \leqq p^2$. On the other hand, if $|C_{A_0}(\overline{B})| \geqq p^2$, we would have a non-cyclic subgroup of A which centralizes $\overline{B}$ and therefore B, a contradiction to 7.5. Therefore $|A_0| = p^3$. If A_0 centralizes $Z(B)$, then A_0 will induce a symplectic group of automorphisms of $B/Z(B)$ and so a subgroup V of A of type (p, p) will centralize $B/Z(B)$ and $Z(B)$. It follows that V centralizes B, a contradiction to 7.5.

Let $D = C_{A_0}(Z(B))$. Since $A_0/D \subseteqq \text{aut } Z(B)$, A_0/D is cyclic and so $|D| = p^2$. Since $C_B(A_0) \subseteqq C_B(D) = Z(B)$ and $[A_0, Z(B)] \neq 1$, we have $C_B(A_0) = 1$. Therefore, in this case, $E = Z(B) \trianglelefteq X$ and so $N_G(E) = X$. Finally since $D/C_B(\overline{B})$ is cyclic, there exists an element $d \in D$ such that $B \subseteqq C_G(d)$. Then $C_G(d) \subseteqq X$, a uniqueness subgroup for B. Thus the result is completely proved if B is non-abelian of order q^3.

For the remainder of the proof, we assume that B is not non-abelian of order q^3. Thus X is a uniqueness subgroup for every subgroup of B of order q^2.

Now the image of any irreducible representation of A on a vector space of characteristic $q \neq p$ is cyclic. Thus each minimal A-invariant subgroup of $Z(B)$ or $B/Z(B)$ is cyclic. Otherwise A_0 has a subgroup of order $\geqq p^2$ which centralizes a subgroup of order q^2 of B, contradicting 6.5. Remember that the fixed points of A on $B/Z(B)$ are just images of fixed points of A by 0.3. Let N be an A-invariant subgroup of B of order q^2. If $|Z(B)| \geqq q^2$ we may choose $N \subseteqq Z(B)$ by Maschke's Theorem, while if $|Z(B)| = q$ we may choose $N/Z(B)$ a minimal A-invariant subgroup of $B/Z(B)$.

As before $A_0/C_{A_0}(N) \subseteqq GL(2, q)$ and so $|A_0/C_{A_0}(N)| \leqq p^2$. By 7.5 $|C_{Z_0}(N)| \leqq p$ and so $|A_0| = p^3$. Let N_1 be a minimal A-invariant subgroup of N and define $D = C_{A_0}(N_1)$. We claim that D satisfies the conclusion of the Lemma.

First if A_0 centralizes N_1, A_0 induces a cyclic group of automorphisms on N/N_1. Thus a subgroup of order p^2 of A_0 centralizes N/N_1 and N_1. This contradicts 7.5. Therefore $D \neq A_0$ and $C_B(A_0) \subseteqq C_B(D) \cap C(A_0) = N_1 \cap C_B(A_0) = 1$. Since $A_0/D \subseteqq \text{aut } N_1$, $|D| = p^2$. In this case $E = N_1$ and $N_G(E) \supseteqq N$, while X is a uniqueness subgroup for N by 7.2. Thus $N_G(E) \subseteqq X$. Finally since D centralizes N_1 and induces a cyclic group of automorphisms of N/N_1 some element $d \in D$ centralizes N/N_1 and N_1. Then $N \subseteqq C_G(d)$. But X is a uniqueness subgroup for N by 7.2. Thus $C_G(d) \subseteqq X$.

This completes the proof. //

Lemma 7.7. _If_ R _is any_ p-_subgroup of_ X _containing_ A, _then_ $N_G(R) \subseteq X$.

Proof. Let Q^* be a maximal element of $\mathcal{U}_G(A, q)$ containing B. Then $Q^* \subseteq X$, because X is a uniqueness subgroup for B. By the Transitivity Theorem 4.3, every element of $\mathcal{U}_G(A, q)$ lies in X.

Note that $C_G(A) \subseteq C_G(D) \subseteq X$ by 7.6(iii). Now if R is a p-subgroup of X containing A, then $\langle \mathcal{U}_G(R, q)\rangle \subseteq X$ and $B \subseteq \langle \mathcal{U}_G(R, q)\rangle$. As $N_G(R)$ normalizes this subgroup $\langle \mathcal{U}_G(R, q)\rangle$, we have $N_G(R) \subseteq X$ since X is a uniqueness subgroup for B. This completes the proof. //

We are now in a position to contradict the existence of G. We know by 7.1 that H is a group of uniqueness type and so the results of 6.4, 7.5, 7.6, 7.7 apply to it.

Lemma 7.8. _Put_ $X = H$ _and let_ E, D, B _be as defined in_ 7.6. _If_ B _is not non-abelian of order_ q^3, _and_ $S = P \cap O_{p',p}(H)$, _then_ $N_H(S) \subseteq N_G(E)$.

Proof. By 5.6, if $C_G(d) \subseteq H$ for all $d \in D^{\#}$, H would be a uniqueness subgroup for D and A. Thus there exists an element $d \in D^{\#}$ such that $C_G(d) \nsubseteq H$. Let K be a maximal subgroup of G containing $C_G(d)$. Then $K = H$ while $E = C_B(D) \subseteq C_G(d) \subseteq K$.

If K is of uniqueness type, since $A \subseteq K$, we may apply 7.7 and get that $P \subseteq K$. But A is an abelian normal subgroup of P and since H is p-stable, $A \subseteq S$. By 7.7 again applied to K, $N_G(S) \subseteq K$. Thus $N_G(S) \cap H$ normalizes $B \cap K \supseteq C_B(D)$. If $B \cap K \neq C_B(D)$, since H is a uniqueness subgroup for all subgroups of B of type (p, p) by 7.2, it follows that $H = K$, a contradiction. Remember we have assumed B is not non-abelian of order q^3. Thus $B \cap K = C_B(D)$ is normalized by $N_G(S)$.

Hence we may assume that K is not of uniqueness type. We derive a contradiction.

We show first that E centralizes $F(K)_{q'}$. For if r is a prime different from q such that $[E, F(K)_r] \neq 1$, we note that $F(K)_r$ has no elementary abelian subgroup of order r^3. Since A normalizes E, we may consider the action of EA on $F(K)_r$. By a now hopefully familiar argument, using a short trip via the Thompson subgroup, we see that a homomorphic image $\overline{EA}$ of EA is a subgroup of $GL(2, r)$. But then $\overline{EA}$, a subgroup of odd order of $GL(2, r)$ is abelian, since the derived subgroup of any such subgroup if an r-group and $\overline{E} \lhd \overline{EA}$. Hence $[E, A] \subseteqq E \cap C_G(F(K)_r)$. Thus either $E \subseteqq C_G(F(K)_r)$ or $[E, A] = 1$. This last possibility does not occur by 7.6 since $C_B(A_0) = 1$. Thus E centralizes $F(K)_{q'}$.

Now if $F(K)_{q'} = F(K)$, we have a contradiction because then $E \subseteqq C_G(F(K)) \subseteqq F(K)$.

Since $N_G(E) \subseteqq H$ by 7.6, $E \ntrianglelefteq K$. Thus $EF(K)_q \cap H \supseteqq E(C_G(E) \cap F(K)_q) \supset E$, since $|E| = q$, and $EF(K)_q \cap H$, being non-cyclic and of odd order contains an elementary abelian q-subgroup Q of order q^2. Now we have just shown that $[E, F(K)_{q'}] = 1$, while $[F(K)_q, F(K)_{q'}] = 1$ obviously. Thus $[Q, F(K)_{q'}] = 1$. Now $F(K)_{q'} \subseteqq C_G(E) \subseteqq H$. If some element of order q in $C_H(Q)$ did not lie in Q we would have Q contained in a subgroup of type (q, q, q) of H and by 7.4, H would be a uniqueness subgroup for Q. This is not the case since $Q \subseteqq K$. We thus have $F(K)_{q'} \subseteqq H$ normalizes QB and centralizes the subgroup Q which contains all elements of order q in its centralizer. Apply 2.2 and see that $[F(K)_{q'}, B] = 1$. Thus $N(F(K)_{q'}) \supseteqq B$. But H is a uniqueness subgroup for B. Hence $K = N(F(K)_{q'}) \subseteqq H$ unless $F(K)_{q'} = 1$ and $F(K)$ is a q-group. But then $F(K)$ has no subgroup of type (q, q, q). Let C be a Thompson subgroup of $F(K)_q$. Then $C_K(\overline{C})$ is a q-group and $K/C_K(\overline{C}) \subseteqq GL(2, q)$ as usual, where $\overline{C} = C/\Phi(C)$. Thus K has a normal Sylow q-subgroup $F(K)$. But $A \subseteqq K$ acts faithfully on $F(K)$. This is impossible since A is a 3-generated group and $F(K)$ has no subgroup of type (q, q, q).

This completes the proof. //

Lemma 7.9. <u>Without loss of generality there exists</u> $z \in Z(P)$ <u>such that</u> $C_G(z) \subseteqq H$.

Proof. Let $B = \Omega_1(Z_2(F(H)_q))$ as usual. If B is non-abelian of order q^3, then $C_A(B) \neq 1$ since A has a subgroup of type (p, p, p). Also $C_P(B) \trianglelefteq P$. If P is abelian, choose $z \in C_A(B)$. Then $C_G(z) \supseteqq B$ and so $C_G(z) \subseteqq H$, a uniqueness subgroup for B. If P is non-abelian choose $z \in P' \cap Z(P)$. Since $P/C_P(B)$ is a p-subgroup of $GL(2, q)$ where p is odd, $P/C_P(B)$ is abelian. Thus $z \in C_P(B)$ and so $C_G(z) \supseteq B$, whence $C_G(z) \subseteq H$. Hence if B is non-abelian of order q^3 we are done.

We may therefore apply 7.8. We then have that P normalizes $E = C_B(D)$ where $|E| = q$. Thus P' centralizes E. Now $[A_0, E] \neq 1$ by 7.6 and so $|C_{A_0}(E)| = p^2$. Here $A_0 = \Omega_1(A)$ as usual. By 7.6, $|A_0| = p^3$.

If P is abelian then we may choose $z = d$ in 7.6 and the result holds. If $P' \cap Z(P)$ is non-cyclic, then, since by 7.5 the centralizer of any 2-generated subgroup of A in B has order at most q, we see that $C_B(P' \cap Z(P)) = E$. Then we may choose $D \subseteq P' \cap Z(P)$ without any loss and then for some $d \in D$ we have $C_G(d) \subseteq H$. We are just left with the case $P \cap Z(P)$ is cyclic. But H was chosen in the beginning of §7 as a maximal subgroup containing $N(Z(P) \cap P') \supseteq N_G(P)$ if P is non-abelian. Thus if $z \in Z(P) \cap P'$, $C_G(z) \subseteq H$. This completes the proof. //

The proof of Theorem 5.1 now proceeds very quickly.

Lemma 7.10. <u>If</u> $A \subseteq H \cap H^g$, <u>then</u> $g \in H$.

Proof. Let R be a Sylow p-subgroup of $H \cap H^g$ containing A. Then $R \neq P$, because if $P \subseteq H \cap H^g$, we have $P, P^{g^{-1}} \subseteq H$. Thus $P = P^{g^{-1}h}$ for some $h \in H$ and so $g^{-1}h \in N_G(P) \subseteq H$. Therefore $g \in H$, a contradiction. Consequently $N_G(R) \nsubseteq H$. But an application of 7.7 shows that $N_G(R) \subseteq H$. //

Lemma 7.11. <u>If</u> $z \in P$ <u>is such that</u> $C_G(z) \subseteq H$, <u>then</u> z <u>lies in a unique conjugate of</u> H <u>in</u> G.

Proof. Among all $g \in G$ such that $z \in H \cap H^g$, $g \notin H$ choose g such that $|H \cap H^g|_p$ is maximal. Let S be a Sylow p-subgroup of $H \cap H^g$ containing z. We may clearly assume without any loss that $S \subseteq P$. Also $S \neq P$ since H contains $N_G(P)$. Let $S \subset T \subseteq P$ be a Sylow p-subgroup of $N_H(S)$. Since $T \supset S$, T must be a Sylow p-subgroup of $N_G(S)$. For otherwise we could find a p-subgroup $T_1 \triangleright T$ such that $T_1 \nsubseteq H$ and then if $x \in T_1 - H$ we would have $T \subseteq H \cap H^x$

while $T \supset S$. Let $U = T \cap O_{p',p}(N_G(S))$.

Now $N_G(S) \nsubseteq H$ since $S \neq P$. By 7.10 it follows that $A \nsubseteq S$. Thus $SN_A(S) \supset S$. If $n \in N_A(S)$, we have $[S, n, n] = 1$ and since $N_G(S)$ is p-stable, $n \in O_p(N(U) \cap N_G(S) \bmod C_G(U) \cap N_G(S))$. But $C_G(U) \cap N_G(S) \subseteq O_{p',p}(N_G(S))$ because $N_G(S)$ is p-constrained. Thus $N_A(S) \subseteq O_{p',p}(N_G(S)) \cap P = U$. It follows that $U \supset S$ and so $N_G(U) \subseteq H$, by the maximality of S.

But $N_G(S) = O_{p'}(N_G(S))(N(U) \cap N_G(S))$. Since $O_{p'}(N_G(S)) \subseteq C_G(S) \subseteq C_G(z) \subseteq H$ we have the desired contradiction, viz. $N_G(S) \subseteq H$. This completes the proof. //

Lemma 7.12. H/H' is a p'-group.

Proof. Suppose that $a \in P$ and that $a^g \in P$ for some $g \in G$. Then if $z \in Z(P)$ is chosen so that $C_G(z) \subseteq H$ by Lemma 7.9, we have $z, z^{g^{-1}} \in C_G(a)$. There exists $x \in C_G(a)$ such that $\langle z, z^{g^{-1}x}\rangle$ is a p-group. Thus we can find $y \in G$ such that $\langle z, z^{g^{-1}x}\rangle^y \subseteq P$. But then, $z, z^y \in P$ implies that $y \in H$ by Lemma 7.11. Thus $z^{g^{-1}x} \in H$ and so $g^{-1}x \in H$. Then $a^g = a^{x^{-1}g}$ and $x^{-1}g \in H$. Thus the largest p-factor group of G is isomorphic to that of H by an application of the transfer. Since G is simple we must have H/H' a p'-group. This completes the proof of Lemma 7.12. //

Lemma 7.13. $p \,|\, |H/H'|$. Hence H is a uniqueness subgroup for A.

Proof. For let $S = P \cap O_{p',p}(H)$. By Lemma 7.8, $N_H(S)$ either normalizes E or $|B| = |\Omega_1(Z_2(F(H)_q))| = q^3$. If $N_H(S) \subseteq N_G(E)$ then $N_G(S)$ has a non-trivial p-factor group because $N_G(E)$ does. Remember that A does not centralize E by 7.6(i). This is impossible because $H = O_{p'}(H)N_H(S)$. On the other hand, if $|B| = q^3$, letting $\overline{B} = B/Z(B)$ we have $H/C_H(\overline{B}) \subseteq GL(2, q)$. Again A does not centralize $\overline{B}$ and so $H/C_H(\overline{B})$ is not a p'-group. But as a subgroup of $GL(2, q)$ with $p \neq q$ we have $p \,|\, |H/H'|$. //

8. THE BURNSIDE $p^a q^b$-THEOREM, p, q ODD

This theorem appears in [10] and the proof is due to Goldschmidt. However it owes a great deal to the work of Bender, since it relies on ideas developed by him in the proof of the Uniqueness Theorem. Of course, the theorems in §5, §6, §7 are obviously directly applicable since they are concerned with the structure of a minimal simple group of odd order.

Theorem 8. 1. *If* G *is a group of order* $p^a q^b$ *where* p, q *are odd primes, then* G *is solvable.*

Proof. Let G be a minimal counter example. Then G is obviously a minimal simple group of odd order. Let P be a Sylow p-subgroup, Q a Sylow q-subgroup of G.

(i) $\mathcal{M}_G(R, r') = \{1\}$ for any Sylow r-subgroup of G.

For if $X \in \mathcal{M}_G(R, r')$, there exists a Sylow r'-subgroup $R' \supseteqq X$. Then

$$X^G = X^{RR'} = X^{R'} \subseteqq R'$$

and X^G is a proper normal subgroup of G.

Let H be a maximal subgroup of G.

(ii) If $|\pi(F(H))| = 2$, then $F(H)$ is non-cyclic.

For if $r = \max(p, q)$, then $H = C_H(F(H)_{r'})$ clearly and $C_G(F(H)_r) = C_G(F(H)) \subseteqq F(H)$, assuming $F(H)$ is cyclic.

Thus a Sylow r-subgroup of H is either cyclic and contained in $F(H)$ or non-abelian and metacyclic. In either case $H = N_G(Z)$ where $Z = \Omega_1(F(H)_r)$. But Z char H_r and so a Sylow r-subgroup of H is a Sylow subgroup of G. This contradicts (i).

(iii) $F(H)$ is a r-group for all maximal subgroups H.

By 5. 4, H is the only maximal subgroup of G which contains $Z(F(H))$. Let $V \subseteqq F(H)$ be a group of type (r, r). Then if $x \in V^{\#}$, $C_G(x) \supseteqq Z(F(H))$ and so $C_G(x) \subseteqq H$. By 5. 5, H is only maximal subgroup of G which contains V. Then clearly H contains a Sylow r-subgroup of G, a contradiction to (i).

(iv) Every Sylow r-subgroup R of G lies in a unique maximal subgroup M of G and every maximal subgroup of G contains a Sylow r-subgroup of G for some r.

For $M = N_G(Z(J(R)))$ since F(M) is a r-group for some prime r.

(v) Let R be a Sylow r-subgroup of G, M a uniqueness subgroup containing R. Then M is a uniqueness subgroup for Z(R).

For suppose $Z(R) \subseteqq M_1 \neq M$, M_1 maximal in G. Choose M_1 such that $|M_1 \cap M|_r$ is maximal. Let R_1 be a Sylow r-subgroup of $M_1 \cap M$, $R_1 \supseteqq Z(R)$. Conjugating R_1 by a suitable element of M we may suppose $R_1 \subseteqq R$. By (iv) $R_1 \neq R$. Since $|N_G(R_1)| > |R_1|$, the choice of M_1 ensures that $N_G(R_1) \subseteqq M$. Hence R_1 is a Sylow r-subgroup of M_1. But then M_1 contains a Sylow r'-subgroup R', by (iv). Then if $g \in G$, $M_1^g = M_1^x$, $x \in R$ since $G = R'R$, $R' \subseteqq M_1$.

Thus $Z(R) \subseteqq \bigcap_{g \in G} M_1^g \lhd G$.

(iv) There exist R_1, R_2 Sylow r-subgroups of G such that $R_1 \cap R_2 = 1$.

For let M be the uniqueness subgroup for R_1, $Z(R_1)$. Choose $R_2 \nsubseteq M$ such that $|R_2 \cap M|$ is maximal. Let $R_2 \cap M = S$. Without loss, $S \subseteqq R_1$. Hence $N_G(S) \supseteqq Z(R_1)$ and so $N_G(S) \subseteqq M$ if $S \neq 1$. But $S \subset R_1$, $S \subset R_2$ clearly shows that $N_G(S) \nsubseteq M$.

Theorem 8.1 now follows immediately. For we can choose $r^c = \max(p^a, q^b)$. Then there exists Sylow r-subgroups R_1, R_2 such that $R_1 \cap R_2 = 1$. Then $|G| \geqq |R_1||R_2| = r^{2c} > p^a q^b = |G|$, a contradiction! //

9. MATSUYAMA'S PROOF OF THE $p^a q^b$-THEOREM, p = 2

Lemma 9.1. If G is a p-group and $H \subseteqq G$, then either $H \trianglelefteqq G$ or $N_G(H) \supseteqq H^x \neq H$, $x \in G$.

Proof. The group H acts on the set S of all conjugates $H^x \neq H$ for some $x \in G$ by conjugation. If $S = \emptyset$, then $H \trianglelefteqq G$. Since $|\{H\} \cup S| = [G : N_G(H)] \neq 1$ if $H \ntrianglelefteq G$, $p \nmid |S|$. Therefore there exists $H^x \neq H$ such that $H^x \subseteqq N_G(H)$. //

Theorem 9.2 [18]. If G is a group of order $2^a p^b$, then G is solvable.

Proof. Let G be a minimal counter example. Clearly G is a minimal simple group. By the proof of step (i) in 8.1, no Sylow r-subgroup of G normalizes any non-trivial r'-subgroup.

(i) If H is a maximal subgroup of G such that $\pi(F(H)) = \{2, p\}$, then $F(H)$ contains a subgroup of type (r, r) for some prime r.

For otherwise $F(H)_2$ is cyclic or quaternion while $F(H)_p$ is cyclic. If $F(H)_2$ is cyclic, then $H_p \subseteq C_H(F(H)_2)$ and so $Z(H_p) \subseteq C_H(F(H)) \subseteq F(H)$. Thus $Z(H)_p \lhd H$ and $Z(H_p)$ char H_p. It follows that H_p is a Sylow p-subgroup of G, a contradiction.

If $F(H)_2$ is quaternion, then since $H/C_G(F(H)_p)$ is abelian, $H'_2 \subseteq C_G(F(H)_p)$. Thus $Z(H_2) \cap H'_2 \subseteq C_G(F(H)) \subseteq F(H)$. Let Z be the unique subgroup of order 2 in $F(H)$. Then Z char $Z(H_2) \cap H'_2$ char H_2. Since $H = N_G(Z)$, we again have H_2 is a 2-subgroup of G, a contradiction.

(ii) If H is a maximal subgroup of G such that $\pi(F(H)) = \{2, p\}$, then H is the only maximal subgroup of G containing $Z(F(H))$.

For let $K \supseteq Z$ be maximal in G, $Z = Z(F(H))$. First $Z_p \subseteq O_p(N_K(Z_2))$ and so $Z_p \subseteq O_p(K)$ by 3.2. Thus $[Z_p, F(K)_2] = 1$. Hence $F(K)_2 \subseteq N_G(Z_p) = H$. Similarly $F(K)_p \subseteq H$.

Now $F(K)_2 \subseteq O_2(N_H(F(K)_p)) \subseteq F(H)_2$ by 3.2 again. By symmetry, $F(H)_2 \subseteq F(K)_2$ and $H = K$.

(iii) If H is a maximal subgroup of G, then $F(H)$ is an r-subgroup for some prime r.

Let $V \subseteq F(H)$ be of type (r, r), using step (i).

Then if $x \in V^{\#}$, $C_G(x) \supseteq Z(F(H))$ and so $C_G(x) \subseteq H$ by (ii). Let $R \supseteq V$ be a Sylow r-subgroup of H. Let S be a R-invariant r'-subgroup of G. Then $S = \langle C_S(x) : x \in V^{\#}\rangle$. Thus $S \subseteq H$. Since $H = RS_1$ where S_1 is a Sylow r'-subgroup of H containing S, $S^H = S^{RS_1} = S^{S_1} \subseteq S_1$ is a normal r'-subgroup of H. Thus $S \subseteq F(H)_{r'}$. Hence $F(H)_{r'}$ is the unique maximal R-invariant r'-subgroup of G. If $R \subseteq R_1$ where R_1 is a Sylow r-subgroup of G, then $N_{R_1}(R)$ normalizes $F(H)_{r'}$ and so lies in H. Hence $R = R_1$ is a Sylow subgroup of G, a contradiction.

(iv) There exists a maximal subgroup M such that $M \cap Z(P) \neq 1$, $M \cap Z(Q) \neq 1$ where P is a Sylow p-subgroup of G and Q is a Sylow 2-subgroup of G.

For let Z be a conjugacy class of involutions of G containing an element x of Z(Q). There exists $y \in Z$ such that $\langle x, y\rangle$ is not a 2-group by 1.1. Thus xy has order divisible by p. Let $H \subseteq \langle xy\rangle$ be the unique subgroup of order p in $\langle xy\rangle$. Choose a maximal subgroup M of G containing $N_G(H)$. Clearly $M \cap Z(Q) \neq 1$, while just as obviously $Z(P) \subseteq N_G(H) \subseteq M$, where P is a Sylow p-subgroup containing H.

We may now complete the proof of 9.2. Let M be a maximal subgroup of G satisfying the conditions of (iv). By (iii), F(M) is a r-group. Let $R \supseteq M_r$ be a Sylow r-subgroup of G and let $S \supseteq M_{r'}$ be a Sylow r'-subgroup of G. First $Z(R) \subseteq C_G(F(M)_r) \subseteq F(M)$. Choose $x \in M \cap Z(S)$ and let $N_0 = \langle Z(R)^{x^i} : i \in Z\rangle$. Clearly $N_0 \subseteq F(M)$ is an r-group. Let $\Omega = \{Z(R)^y : y \in G\} = \Omega_1 \cup \Omega_2 \cup \ldots \cup \Omega_s$, where Ω_1 is an $\langle x\rangle$ orbit in Ω.

Let $N_i = \langle \Omega_i \rangle$. Since $G = RS$, there exists $y \in S$ such that $Z(R)^y \in \Omega_i$ and so $N_i = \langle Z(R)^{yx^i} : i \in Z\rangle = \langle Z(R)^{x^i y} : i \in Z\rangle = N_0^y$.

Thus N_i is an r-group normalized by $\langle x\rangle$. Choose l maximal such that $N = \langle \Omega_{i_1}, \ldots, \Omega_{i_l}\rangle$ is an r-group normalized by $\langle x\rangle$. Assume for simplicity that $N = \langle \Omega_1, \ldots, \Omega_l\rangle$. Of course $x \in N_G(N)$. Let T be a Sylow r-subgroup of G containing N. By 9.1, either $N \trianglelefteq T$ or there exists $N^z \neq N$ such that $N^z \subseteq N_G(N)$, $z \in T$.

If $N \trianglelefteq T$ then since $G = ST$, any G-conjugate of $x \in Z(S)$ is a T-conjugate and so $\langle x^G\rangle \subseteq N_G(N) \subset G$, a contradiction.

If $N_G(N) \supseteq N^z \neq N$, $z \in T$, then since $\Omega_1^z \cup \ldots \cup \Omega_l^z \nsubseteq N$, there exists $s \in S$ such that $Z(R)^s \subseteq N^z$ and $Z(R)^s \nsubseteq N$. Suppose $Z(R)^s \in \Omega_i$, $i > l$. Then $N_i \subseteq N_G(N)$ ($\ni x$) and so $N_G(N) \supseteq Z(R)^s$. Now NN_i is an r-group normalized by x and generated by $\Omega_1, \ldots, \Omega_l, \Omega_i$, a contradiction. This completes the proof. //

10. A GENERALIZATION OF THE FITTING SUBGROUP

Definition. Let X be any group, $F(X)$, the Fitting subgroup of X. We define $F^*(X) = \text{socle}\ (F(X).C_X(F(X)) \bmod F(X))$. Define $E(X)$ to be the terminal member of the derived series of $F^*(X)$.

It is easy to see that $F(X)C_X(F(X))/F(X)$ has no solvable normal subgroup. For then we could choose a p-group $P \subseteq C_X(F(X))$ such that $PF(X)/F(X)$ is minimal normal in $X/F(X)$ and then $PF(X)$ is a nilpotent normal subgroup of X. Thus $F(X)C_X(F(X))/F(X)$ has no solvable normal subgroup and its socle is a direct product of non-abelian simple groups. It is easy to see that $F^*(X) = F(X)E(X)$ and that $C_X(F^*(X)) \subseteq F^*(X)$. Since this is actually the most important property of the group $F^*(X)$ - being easily true when X is solvable - we verify this in the following

Lemma 10.1. (a) $F^*(X) = F(X)E(X)$.

(b) $[F(X), E(X)] = 1$.

(c) $C_X(F^*(X)) \subseteq F^*(X)$.

Proof. (a) is clear.

(b) $F(X)C_X(F(X))/C_X(F(X))$, being a homomorphic image of a nilpotent group is solvable. Thus $E(X) = (F^*(X))^{\infty} \subseteq (F(X)C_X(F(X)))^{\infty} \subseteq C_X(F(X))$.

(c) Suppose $C_X(F^*(X)) \nsubseteq F^*(X)$. Then $C_X(F^*(X))F(X)/F(X) \trianglelefteq C_X(F(X))F(X)/F(X)$ and $C_X(F^*(X)) \cap F^*(X) = Z(F^*(X)) \subseteq F(X)$. Thus $C_X(F^*(X))F(X)/F(X) \cap F^*(X)/F(X) = 1$. We see therefore that there exists a minimal normal subgroup of $F(X)C_X(F(X))/F(X)$ which avoids $F^*(X)/F(X)$ and this is clearly impossible since $F^*(X) = \text{socle}\ (F(X)C_X(F(X)) \bmod F(X))$. This completes the proof. //

Since by 7.1(b), $[F(X), E(X)] = 1$, we have $F(X) \cap E(X) \subseteq Z(E(X))$. Also $E(X)/F(X) \cap E(X) \cong F^*(X)/F(X)$, a direct product of non-abelian simple groups. Thus $Z(E(X)) = F(X) \cap E(X)$. Now $E(X)/Z(E(X))$ is a direct product of non-abelian simple groups $S_i/Z(E(X))$, $1 \leq i \leq n$. Define $E_i = S_i^{(\infty)}$. The groups E_i are quasi-simple - that is $E_i/Z(E_i)$ is a non-abelian simple group. They are called the <u>components</u> of X. We will frequently write $\overline{E}_i$ for $E_i/Z(E_i)$.

Lemma 10.2. (a) $[E_i, E_j] = 1$ if $i \neq j$.

(b) $E(X) = E_1 \ldots E_n$.

Proof. (a) $[E_i, E_j] \subseteq Z(E(X))$ since $E(X)/Z(E(X))$ is a direct product of groups $E_iZ(E(X))/Z(E(X))$. Thus

$$[E_i, E_j, E_i] = 1 \text{ if } i \neq j.$$

It follows by the Three Subgroups Lemma that $[E_i, E_i, E_j] = 1$. Since E_i is perfect we have $[E_i, E_j] = 1$.

(b) Clearly $E(X) = E_1 \ldots E_nZ(E(X))$.

Hence $E(X) = E(X)' = (E_1 \ldots E_n)' = E_1 \ldots E_n$. This completes the proof. //

The most remarkable property of the groups E_i is contained in the next Lemma. We see there that any E_i-invariant solvable subgroups of X(!) must be actually centralized by E_i.

Lemma 10.3. (a) $C_X(E_i \bmod Z(E(X))) = C_X(E_i)$ <u>for any component</u> E_i <u>of</u> X.

(b) <u>Any</u> E_i<u>-invariant solvable subgroup</u> S <u>of</u> X <u>is centralized by</u> E_i.

Proof. (a) Let $x \in C_X(E_i \bmod Z(E(X)))$. Then $[E_i, x] \subseteq Z(E(X))$ and so $[E_i, x, E_i] = 1$.

The Three Subgroups Lemma gives $[E_i, E_i, x] = 1$ and the perfectness of E_i shows that $[E_i, x] = 1$. Thus $C_X(E_i \bmod Z(E(X))) \subseteq C_X(E_i)$. The other containment is obvious.

(b) Let S be a solvable E_i-invariant subgroup of X. Then $[E_i, S] \subseteq E(X) \cap S$. Now $E(X) \cap S$ is a solvable E_i-invariant subgroup of $E(X)$. Consider the image of $E(X) \cap S$ in $E(X)/Z(E(X))$, a direct product of non-abelian simple groups $\overline{E}_1 \times \ldots \times \overline{E}_n$. Consider the projection maps $\pi_j : E(X)/Z(E(X)) \to \overline{E}_j$. Then $\pi_j((E(X) \cap S)Z(E(X))/Z(E(X)))$ certainly commutes with $\overline{E}_i$, if $j \neq i$.

On the other hand $\pi_i((E(X) \cap S)Z(E(X))/Z(E(X)))$ is a solvable normal subgroup of $\overline{E}_i$, a non-abelian simple group. Thus $(E(X) \cap S)Z(E(X))/Z(E(X))$ centralizes $\overline{E}_i$ and so $E(X) \cap S$ commutes with E_i modulo $Z(E(X))$. By 7.3(a), $E(X) \cap S$ actually centralizes E_i.

Thus $[E_i, S, E_i] = 1$. The Three Subgroups Lemma with $E_i' = E_i$ shows that $[E_i, S] = 1$. This completes the proof. //

Considerable interest will be attached to the subnormal subgroups of $F^*(X)$. The following Lemma indicates part of their structure.

Lemma 10.4. _Suppose that_ $S \lhd\lhd F^*(X)$. _Then_

(a) $F(S) = F(X) \cap S$

(b) $S = (F(X) \cap S)(E(X) \cap S) = F(S)E(S)$.

Proof. First $F(X) \cap S \subseteqq F(S)$. Let $S = S_0 \lhd \dots \lhd S_n = F^*(X)$. Then by induction $F(S) \subseteqq F(S_i)$ for all $i = 1, 2, \dots, n$ and so $F(S) \subseteqq F(X) \cap S$. This proves (a).

Hence $S/F(S) = S/F(X) \cap S = SF(X)/F(X)$ is a subnormal subgroup of $F^*(X)/F(X)$ and so is a direct product of non-abelian simple groups or the identity group. Hence $S/F(S) = (S/F(S))^{(\infty)} = S^{(\infty)}F(S)/F(S)$ and so $S = S^{(\infty)}F(S)$. Then $E(S) \subseteqq S^{(\infty)} \subseteqq F^*(X)^{(\infty)} \cap S = E(X) \cap S$. Let $^-$ denote the natural homomorphism modulo $F(S)$. Then we have $\overline{E(X) \cap S} \subseteqq C_{\overline{S}}(\overline{F(S)})$ and $\overline{E(X) \cap S}$ is a normal subgroup of $\overline{S}$, a direct product of simple groups. Therefore $\overline{E(X) \cap S} \subseteqq \overline{E(S)}$ and we have $E(S) \subseteqq S^{(\infty)} \subseteqq E(X) \cap S \subseteqq E(S)F(S)$. It follows that $S = E(S)F(S) = (E(X) \cap S)(F(X) \cap S)$. This completes the proof. //

Lemma 10.5. _Suppose_ $S \lhd\lhd F^*(X)$. _Then_

(a) $N_{F^*(X)}(S) \lhd\lhd F^*(X)$

(b) _If_ $C_{F^*(X)}(S) \subseteqq S$, _then_ $E(S) = E(X)$.

Proof. Apply 10.4 and get $S = E(S)F(S) = (E(X) \cap S)(F(X) \cap S)$. Clearly $E(X) \subseteqq N_X(E(X) \cap S) \cap C_X(F(X) \cap S)$. Thus $E(X) \subseteqq N_X(S)$ and (a) follows because $F(X)$ is nilpotent.

To verify (b) we show that $E(X) \subseteqq S$ and the result follows from 10.4. Suppose that E_1 is a component of $E(X)$ not contained in S. Then if $\overline{E}_1 = E_1F(X)/F(X)$ etc., we have $[\overline{E}_1, \overline{S}] = 1$. Hence $[E_1, S] \subseteqq F(X) \cap E(X) = Z(E(X))$. We may apply Lemma 10.3(a) and get $[E_1, S] = 1$. This contradicts $C_{F^*(X)}(S) \subseteqq S$. //

Lemma 10.6. Suppose that $A \lhd\lhd X$ and $A = E(A)$. Then $A \trianglelefteq E(X)$.

Proof. Suppose that $A = A_0 \lhd \ldots \lhd A_n \lhd A_{n+1} = X$. If $n = 0$ the result is clearly true since $[A, F(X)] \subseteq F(X) \cap A \subseteq Z(A)$ and so $[A, F(X)] = 1$ by a familiar argument.

Suppose $A \subseteq E(A_n) \trianglelefteq X$. Then $[E(A_n), F(X)] \subseteq F(X) \cap E(A_n) \subseteq Z(E(A_n))$. Thus $[E(A_n), F(X)] = 1$ and $E(A_n) \subseteq E(A_{n+1})$. //

Lemma 10.7. Suppose $A, B \subseteq X$ and $F^*(A) \cap B \lhd\lhd F^*(A)$. Then $F^*(A) \cap B \subseteq F^*(A \cap B)$. Further if $B \trianglelefteq A$ then $F^*(A) \cap B = F^*(B)$.

Proof. Clearly $F(A) \cap B \subseteq F(A \cap B)$. By 10.4,

$$\begin{aligned} F^*(A) \cap B &= (F^*(A) \cap B \cap F(A))(F^*(A) \cap B \cap E(A)) \\ &= (F(A) \cap B)(E(A) \cap B). \end{aligned}$$

Now $E(A) \cap B \trianglelefteq F^*(A) \cap B \lhd\lhd F^*(A)$. Thus $(E(A) \cap B)F(A)/F(A)$ is a product of components of $F^*(A)/F(A)$.

Let E_1 be some component of $F^*(A)$ contained in $(E(A) \cap B)F(A)$. Then $E^1 \subseteq ((E(A) \cap B)F(A))^{(\infty)} \subseteq E(A) \cap B$ and E_1 normalizes $F(A \cap B)$, a solvable subgroup of A. Thus E_1 centralizes $F(A \cap B)$. It follows that $E_1 \subseteq F^*(A \cap B)$.

Now if $B \trianglelefteq A$, then $F^*(A) \cap B \subseteq F^*(B)$. Conversely $F(B) = F(A) \cap B \subseteq F^*(A) \cap B$. Since $E(B) \lhd\lhd A$, by 10.6, $E(B) \subseteq E(A) \cap B$. Hence $F^*(A) \cap B = F^*(B)$. //

Lemma 10.8. Let U be any subgroup of X. Then

$$E(X) = (C_G(U) \cap E(X))[E(X), U].$$

Proof. First $[E(X), U] \trianglelefteq E(X)$. Let E_1 be a component of $E(X)$ not contained in $[E(X), U]$. We show that $E_1 \subseteq C_G(U)$. Since $[E_1, U] \subseteq [E(X), U]$, we have $[E_1, U] \subseteq C_G(E_1)$. Thus $[E_1, U, E_1] = 1$ and, by the Three Subgroups Lemma, $[E_1, U] = 1$. This completes the proof. //

11. GROUPS WITH ABELIAN SYLOW 2-SUBGROUPS

J. H. Walter obtained a characterization of finite groups with abelian Sylow 2-subgroups in [22], [23]. Bender offers a novel approach in [5]. In this section we commence this classification by Bender of groups with abelian Sylow 2-subgroups. This depends on the characterization also due to Bender of groups which have a strongly embedded subgroup. There are many equivalent formulations of this concept. The following definition suffices for our purposes here.

Definition 11.1. A subgroup H of even order of a group G is strongly embedded in G if $H \neq G$ and $H \cap H^x$ has odd order for all $x \in G - H$.

In [4], Bender characterized all finite groups which have a strongly embedded subgroup. If G is a non-abelian simple group and has a strongly embedded subgroup, then $G \cong SL(2, 2^n)$, $Sz(2^{2n+1})$, $U_3(2^n)$ for suitable n. Here $SL(2, 2^n)$ denotes the group of all 2×2 matrices of determinant 1 with coefficients from the field of 2^n elements. The groups $Sz(2^{2n+1})$, $U_3(2^n)$ denote the Suzuki simple groups, see [20], and the projective special unitary 3 dimensional group over a field of 2^{2n} elements, respectively.

We are here interested in groups with abelian Sylow 2-subgroups. Of the above three classes of groups, only the groups $SL(2, 2^n) = L_2(2^n)$ have abelian Sylow 2-subgroups. The known groups which have abelian Sylow 2-subgroups in fact include just three more classes of groups. The projective special linear 2-dimensional groups $L_2(q)$, where q is an odd prime power such that $q \equiv \pm 3 \pmod 8$, all have Sylow 2-subgroups which are elementary abelian of order 4. The simple group J_1 of Janko [16] and the Ree simple groups $G_2^1(q)$, where $q = 3^{2n+1}$, $n \geqq 1$, [19] all have elementary abelian Sylow 2-subgroups of order 8.

It is still unresolved whether this completes the list of groups with abelian Sylow 2-subgroups. The groups of Janko and Ree are clearly the most intriguing groups on the above list. They have a single class of involutions and if t is one such, $C(t)$ is isomorphic to $\langle t \rangle \times E$ where $E \cong L_2(q)$ and q is odd. Of course, it follows that $q \equiv \pm 3 \pmod 8$ since otherwise a Sylow 2-subgroup would be non-abelian. We proceed to define a JR (Janko-Ree) group.

Definition 11. 2. A simple group G with abelian Sylow 2-subgroups is called a JR-group if G contains an involution t such that $C_G(t) = \langle t \rangle \times E$, where $L_2(q) \subseteq E \subseteq P\Gamma L(2, q)$ and q is odd.

It follows that $[E : L_2(q)]$ is odd since otherwise a Sylow 2-subgroup is non-abelian. It is easy to see that a JR-group has a single class of involutions and so the groups J_1, $G_2^1(q)$, $q = 3^{2n+1}$, $n \geqq 1$, are groups of type JR.

It was shown by Walter [22] that in a group of type JR, the group E which occurs as the business end of the centralizer of an involution in fact must be isomorphic to $L_2(q)$. Now a simple group G with an involution t such that $C_G(t) = \langle t \rangle \times E$ where $E = L_2(q)$, $q \equiv \pm 3 (\text{mod } 8)$ first has $q = 5$ in which case $G = J_1$ by [16], or has $q = 3^{2n+1}$, $n \geqq 1$, by [17]. In an early paper, H. N. Ward [24] showed that the character table of G is determined and Janko and Thompson [17], showed that the 3-Sylow normalizer of G has a uniquely determined structure. Further results have been obtained by Thompson [21], but the final determination of the multiplication table of G still eludes us. Thus either a simple group of type JR is J_1 or $G_2^1(q)$ or a new simple group with the same character table (and so order) as $G_2^1(q)$ and with very similar structure.

Definition 11. 3. A group G with an abelian Sylow 2-subgroup is said to be an A*-group, if G has a normal series $1 \subseteq N \subseteq M \subseteq G$ where N and G/M are of odd order and M/N is a direct product of a 2-group and simple groups of type $L_2(q)$ or JR.

Remark. Our definition of a JR-group differs only slightly from that of Bender [5]. His definition requires (and his proof uses crucially) the fact that the centralizer of an involution of a simple group G of type JR is a maximal subgroup of G. This follows from the Definition 11. 2, as will be proved in 12. 2 below.

We state the main theorem of Bender [5] here.

Theorem A. *Let G be a finite group with an abelian Sylow 2-subgroup. Then G is an A*-group.*

Remark. If a group G has an abelian Sylow 2-subgroup T of rank 1, then G is solvable and 2-nilpotent and clearly an A*-group. If T has rank 2, then G is 2-nilpotent unless T is of type $(2^a, 2^a)$. But then if $a > 1$, G is solvable by [7] and clearly an A*-group since it has 2-length 1. Thus we may assume that $|T| = 4$ and then we can apply the results of [13]. Again G is an A*-group.

Hence the main thrust of Theorem A is directed at the case of a finite group G with an abelian Sylow 2-subgroup of rank at least 3.

We will follow here closely the direction of Bender's proof, expanding the abbreviation portions where necessary. Minor changes in the presentation will be made, but the proof itself is so delicately woven and intricate that its beauty would suffer if the changes were too great. For generalizations of the techniques of the following sections, the reader should consult Goldschmidt's 'strongly closed abelian' paper [11].

12. PRELIMINARY LEMMAS

Lemma 12.1 (Thompson). *Let G be a group with an abelian Sylow 2-subgroup S. Let R be a subgroup of S such that $r(R) = r(S) - 1$. If $G = O^2(G)$, then any involution $t \in S$ is conjugate in $N_G(S)$ to an element of R.*

Proof. Of course, $N_G(S)$ controls fusion of elements of S by Burnside's lemma.

Consider the transfer $V : G \to S$.

Let $x_1, \ldots, x_k, x_{k+1}t, \ldots, x_n, x_nt$ be a system of coset representatives of S in G chosen so that $x_itx_i^{-1} \in S$ for $i = 1, \ldots, k$, $x_itx_i^{-1} \notin S$ for $i > k$. Since $[G : S]$ is odd, k is odd.

Clearly $V(t) = \prod_{i=1}^{k} x_itx_i^{-1}$.

If $x_itx_i^{-1}, x_{i+1}tx_{i+1}^{-1} \notin R$, then $x_itx_i^{-1}x_{i+1}tx_{i+1}^{-1} \in R$, since $r(R) = r(S) - 1$. Thus $V(t) \equiv x_1tx_1^{-1} \pmod R$. Since $O^2(G) = 1$, $V(t) = 1$ and so $x_1tx_1^{-1} \in R$. //

Lemma 12.2. Let G be a simple group with an abelian Sylow 2-subgroup S. Suppose $t \in S$ is an involution such that $C_G(t) = \langle t \rangle \times E$ where $L_2(q) \subseteq E \subseteq P\Gamma L(2, q)$. Then $C_G(t)$ is a maximal subgroup of G.

Proof. Suppose $C_G(t) \subseteq M \subset G$.

If $O^2(M) = M$, then by 12.1, t is conjugate in M to an element of E. Since $C_G(S) \subseteq C_G(t) \subseteq M$, and $N_M(S)/C_M(S)$ acts transitively on the non-trivial elements of M, we have $N_M(S) = N_G(S) \subseteq M$. Also for all involutions $s \in M$, $s = t^m$, $m \in M$ and so $C_G(s) = C_G(t^m) \subseteq M$. Suppose $M \cap M^x$ contains an involution s, where $x \in G - M$. Then s, $s^{x^{-1}} \in M$ and so there exists $y \in M$ such that $s^{x^{-1}y} = s$. But then $x^{-1}y \in C_G(s) \subseteq M$, $x \in M$, a contradiction. Hence M is strongly embedded in G.

But by [4], $G \cong L_2(2^n)$ since the Sylow 2-subgroups of G are abelian. But the centralizer of any involution in $L_2(2^n)$ is an abelian 2-group. This contradicts our assumption that $C(t) = \langle t \rangle \times E$ where $E \supseteq L_2(q)$.

If $O^2(M) \subset M$ then let $K = O^2(M)$. Clearly $[M : K] = 2$. Since a Sylow 2-subgroup of $K \supseteq E$ is of order 4, $K \supseteq K_1 \supseteq K_2 \supseteq 1$, K/K_1, K_2 have odd order and $K_1/K_2 \cong L_2(r)$ for some prime power r. Of course $E' \subseteq K_2$. Hence either $q = 5$ or $r = q^k$ for some k. Let s be an involution in K_1. Then $C_{K_1}(s)/C_{K_2}(s)$ is a dihedral group of order $r - \varepsilon$ where $r \equiv \varepsilon (\mathrm{mod}\ 4)$, $\varepsilon = \pm 1$. Since s is conjugate to t and $C_G(t) = \langle t \rangle \times E$, we see that any dihedral section of $C_G(s)$ has order at most $q + 1$. Thus $r - \varepsilon \leqq q + 1$. If $q = 5$ then $r = 5$, while if $q \mid r$ then $r = q$.

If $K_2 \neq 1$, then for some involution $s \in K_1$ we have $C_{K_2}(s) \neq 1$ since there exists a four-subgroup of K_1 normalizing K_2. But then $C_G(s)$ contains a section which has a normal odd order subgroup $C_{K_2}(s)$ with a dihedral factor group of order $q - \varepsilon$. This situation does not arise in $C_G(s) = \langle t \rangle \times E$. Note that any field automorphism of $PSL(2, q)$ acts non-trivially on a dihedral subgroup of order $q - \varepsilon$.

Thus $K_2 = 1$ and $K \subseteq \mathrm{aut}\ L_2(q)$. Since clearly $t \in C_M(K_1) \trianglelefteq M$

we must have $t \in Z(M)$, $M = C_G(t)$. This completes the proof. //

The following lemma is not required for some time in the proof of Theorem A. It is of some independent interest, affording as it does the first glimpse of the importance of the F*-subgroup.

Lemma 12.3. Let G be a group with abelian Sylow 2-subgroups, p a prime, P some p-subgroup of G. Assume that $r(O_p(G)) \leqq 2$ and that $[P, O^p(F^*(G))] = 1$. Then $P \subseteqq O_p(G)$.

Proof. Let $C = C_G(O^p(F^*(G))) \supseteqq P$. If $C \subset G$ then since $C \trianglelefteq G$, by 10.6, $F^*(C) = F^*(G) \cap C$. Because $[P, O^p(F^*(G))] = 1$, $[P, O^p(F^*(C))] = 1$. For $F^*(G) \cap C = (F(G) \cap C)\,(E(G) \cap C)$ and so $O^p(F^*(G) \cap C) = O^p(F(G) \cap C)\, O^p(E(G) \cap C)$. By induction $P \subseteqq O_p(C) \subseteqq O_p(G)$. Thus $C = G$, $E(G) = 1$, $F^*(G) = F(G)$.

Now let $Z = \Omega_1(Z(F(G)_q))$ for some $q \neq p$, and let $\overline{G} = G/Z$. First $F(\overline{G}) = \overline{F(G)}$ because $Z \subseteqq Z(G)$. Also $r(O_p(\overline{G})) \leqq 2$ because $O_p(\overline{G}) = \overline{O_p(G)}$ and Z is a p'-group.

Let $E(\overline{G}) = \overline{E}$. Since $[E^{(\infty)}, F(G), E^{\infty}] = 1$ we have $[E^{(\infty)}, F(G)] = 1$ and because $F^*(G) = F(G)$, $E^{(\infty)} = 1$. Thus $\overline{E} = 1$. Hence $F^*(\overline{G}) = F(\overline{G}) = F^*(\overline{G}) = \overline{F(G)}$. It follows that $[\overline{P}, O^p(F^*(\overline{G}))] = 1$ and by induction $\overline{P} \subseteqq O_p(\overline{G})$. Since $Z \subseteqq Z(G)$, $P \subseteqq O_p(G)$.

Thus $Z = 1$ and $F(G)$ is a p-group containing $C_G(F(G))$. Let Q be a Thompson critical subgroup of $F(G)$. If $p = 2$, then $P \subseteqq C(F(G)) \subseteqq F(G)$ since the Sylow 2-subgroups of G are abelian. If $p \neq 2$, we may choose $R = \Omega_1(Q)$ and then $C_G(R)$ is a p-group by 2.4. Since $r(R) \leqq 2$, $|R| \leqq r^3$ and if $\overline{R} = R/\Phi(R)$, $|\overline{R}| \leqq p^2$. Still $C_G(\overline{R})$ is a p-group and $G/C_G(\overline{R}) \subseteqq GL(2, p)$, with abelian Sylow 2-subgroups. Every such subgroup has a normal Sylow p-subgroup. Thus $PC_G(\overline{R})/C_G(\overline{R}) \subseteqq O_p(G/C_G(\overline{R}))$. Hence $P \subseteqq O_p(G)$. //

The following Theorem is, it seems to me, unique in finite group theory. A theorem about the structure of a general finite simple group with only a very minor restriction on its subgroups! It is followed by an extension very reminiscent of the uniqueness theorems of Chapter 5. It is true to say that these next two Theorems are the very cornerstone of the whole proof.

Theorem B (Bender). Let A and B be a distinct maximal subgroup of a simple group G such that $F^*(A) \subseteq B$ and $F^*(B) \subseteq A$. Then $F^*(A)$ and $F^*(B)$ are p-groups for the same prime p.

Proof. Since $F(B)$ is an $E(A)$-invariant solvable subgroup of A, $F(B) \subseteq C_A(E(A))$ by 10.3.

By 10.8, $E(A) = C_{E(A)}(E(B))[E(A), E(B)]$.

Now $C_{E(A)}(F^*(B)) = C_{E(A)}(E(B)) \subseteq F^*(B)$.

Thus $E(A) \subseteq F^*(B)$ and so $E(A) \subseteq F^*(B)^{(\infty)} = E(B)$.

By symmetry $E(B) \subseteq E(A)$. Since $A \neq B$, $E(A) = E(B) = 1$.

Clearly subgroups of coprime orders of $F(A)$ and $F(B)$ centralize each other. It follows that $\pi(F(A)) = \pi(F(B))$. For otherwise if $p \in \pi(F(A)) - \pi(F(B))$, then $[F(B), O_p(A)] = 1$. Since $F(B) \supseteq C_G(F(B))$, we have a contradiction.

Let $p \in \pi(F(A)) = \pi(F(B))$. Let $P = O_p(A)$, $Q = O_p(B)$, $R = O_{p'}(A)$.

First $[E(R), F(R)] = 1$ and $[O_p(A), E(R)] = 1$. Thus $E(R)$ centralizes $F(R)P = F(A) = F^*(A) \supseteq C_G(F^*(A))$. Hence $E(R) = 1$ and $F^*(R) = F(R)$. Now $[Q, F^*(R)] = [Q, F(R)] = [Q, F(A) \cap R] \subseteq Q \cap R = 1$. Since Q centralizes $F(R) \supseteq C_R(F(R))$, by 2.2, $[Q, R] = 1$. Similarly $[P, O_{p'}(B)] = 1$.

Now $R \subseteq C_G(Q) \trianglelefteq B$. Also $F(B) = QO^p(F(B))$ and $P \subseteq C_B(O^p(F(B))) \trianglelefteq B$. Thus $[P, C_B(Q)] \subseteq C_B(O^p(F(B)) \cap C_B(Q) = C_B(F(B)) \subseteq F(B)$. Because $PF(B) = PQ \times O^p(F(B))$, since subgroups of coprime orders of $F(A)$ and $F(B)$ commute, we have PQ is normalized by $C_B(Q)$. But now

$$[PQ, C_B(Q)] \subseteq [P, C_B(Q)] \subseteq F(B) \cap PQ = Q.$$

Thus $[PQ, C_B(Q), C_B(Q)] = 1$. By 0.2 we have $[PQ, O^p(C_B(Q))]=1$. Hence $O_{p'}(A) \subseteq O_{p'}(O^p(C_B(Q))) \subseteq O_{p'}(B)$.

By symmetry, $O_{p'}(B) \subseteq O_{p'}(A)$.

Hence $O_{p'}(A) = O_{p'}(B) = 1$ and $\pi(F(A)) = \pi(F(B)) = \{p\}$.

This completes the proof of Theorem B. //

Theorem 12.4. Let A be a maximal subgroup of a simple group G, S a subnormal subgroup of $F^*(A)$ such that $C_{F^*(A)}(S) \subseteq S$. Let

$B \subseteq G$ be a subgroup of G containing S. Then

(a) $O_q(B) \cap A = 1$ for $q \in \pi(F(A))'$.

(b) $[O_p(B), O^p(F^*(A))] = 1$, if $p \in \pi(F(A))$.

Moreover, if B is a maximal subgroup of G and if either $|\pi(F^*(A))| \geq 2$ or $|\pi(F^*(B))| \geq 2$, then each of the following ensures that $A = B$.

(c) A contains a subnormal subgroup $\bar{S}$ of $F^*(B)$ such that $C_{F^*(B)}(\bar{S}) \subseteq \bar{S}$.

(d) B is an A^*-group, $|E(B)| \leq |E(A)|$ and $O_q(B) = 1$ for all $q \in \pi(F(A))'$.

(e) A and B are conjugate A^*-groups.

Remark. The structure of A^*-groups is only minimally required in (d) and (e). In fact the relevant fact required of the subgroups A and B is that they have the following structure: $X \supseteq Y \supseteq Z \supseteq 1$ where X/Y, Z are solvable and Y/Z is a direct product of non-abelian simple groups. Thus, roughly speaking, this Theorem holds provided the relevant subgroups do not involve groups of type A_5 wr A_5.

Proof. By 10.5(b), $S \supseteq E(A)$. Also by 10.4, $S = (F(A) \cap S)E(A)$, and since $C_{F^*(A)}(S) \subseteq S$, $Z(F(A)) \subseteq S$. Thus if $p \in \pi(F(A))$, $p \in \pi(F(S))$.

(i) $[O_p(B) \cap A, O^p(S)] = 1$ for all primes p.

For $O_p(B) \cap A$ is an $E(A)$-invariant solvable subgroup of A. Thus

$$[O_p(B) \cap A, E(A)] = 1$$

and

$$[O_p(B) \cap A, O^p(S)] = [O_p(B) \cap A, F(S)_{p'}] \subseteq O_p(B) \cap A \cap F(A)_{p'} = 1.$$

(ii) $C_G(O_q(S)) \cap O_q(A) \subseteq O_q(S)$ for $q \in \pi(F(A))$, $q \neq p$.

For $C_G(O_q(S)) \cap O_q(A)$ centralizes $F(S)_{q'} \subseteq F(A)_{q'}$ and $E(A)$. Thus $C_G(O_q(S)) \cap O_q(A) \subseteq C(S) \cap F^*(A) \subseteq S$. It follows that $C_G(O_q(S)) \cap O_q(A) \subseteq O_q(S) = O_q(A) \cap S$.

(iii) $[O_p(B) \cap A, O_q(A)] = 1$ for $q \in \pi(F(A)) - p$.

This follows from (i) and (ii) using 2.2.

Thus if $p \notin \pi(F(A))$, $O_p(B) \cap A$ centralizes $F^*(A)$ since by (iii) $O_p(B) \cap A \subseteq C_G(F(A))$ and by (ii) $O_p(B) \cap A \subseteq C_G(E(A))$. Since $C_G(F^*(A)) \subseteq F^*(A)$, it follows that $O_p(B) \cap A = 1$. This completes the proof of (a).

Suppose now $p \in \pi(F(A))$.

(iv) $C_G(O_p(S)) \cap O_p(B) \subseteq O_p(B) \cap A$.

For $C_G(O_p(S)) \subseteq C_G(Z(F(A)_p))$ since $C_G(S) \cap F^*(A) \subseteq S$. Thus $C_G(O_p(S)) \subseteq A$.

(v) $O^p(S)$ centralizes $O_p(B)$ and so $[E(A), O_p(B)] = 1$.

For $[O^p(S), O_p(S)] = 1$. By (iv) and (i), $[O^p(S), C(O_p(S)) \cap O_p(B)] = 1$. We may now apply 2.2 to the group $O_p(B)O_p(S)$ to get the result.

(vi) $C(F(S)_{p'}) \cap F(A)_{p'} \subseteq F(S)_{p'}$.

For if $x \in C_G(F(S)_{p'}) \cap F(A)_{p'}$ then $[x, O_p(S)] \subseteq [x, O_p(A)] = 1$. Clearly $[x, E(A)] = 1$ and so $[x, S] = 1$. Since $C_G(S) \cap F^*(A) \subseteq S$, we have (vi).

(vii) We may now apply 2.2 to $F(A)_{p'}$ using (v) and (vi). It follows that $[O_p(B), F(A)_{p'}] = 1$. Hence $[O_p(B), O^p(F^*(A))] = 1$. We are done with (b).

Continuing, let B be a maximal subgroup of G containing S. We show that under condition (c), $F^*(A)$ is a p-group if and only if $F^*(B)$ is a p-group. For if (c) holds, we are assured of a symmetrical relationship between A and B. Thus we have $[O^p(F^*(B)), O_p(A)] = 1$ for all $p \in \pi(F(B))$ and $O_p(A) \cap B = 1$ for all $p \notin \pi(F(B))$.

Now if $F^*(A)$ is a p-group, then since $S \subseteq O_p(A) \cap B$ it follows that $p \in \pi(F(B))$. Thus $O^p(F^*(B)) \subseteq C_G(O_p(A))$. Since $C_G(O_p(A)) \subseteq O_p(A) = F^*(A)$, we see that $F^*(B)$ is a p-group also. Thus $|\pi(F^*(A))| \geq 2$ if and only if $|\pi(F^*(B))| \geq 2$.

Now $O^p(F^*(B)) \subseteq C_G(O_p(A)) \subseteq A$ for all $p \in \pi(F(B))$. Also $[O_p(B), O^p(F^*(A))] = 1$ and so $O_p(B) \subseteq A$ because $O^p(F^*(A)) \neq 1$. Thus $F^*(B) \subseteq A$. Symmetry gives $F^*(B) \subseteq A$ and we may apply Theorem B to get $A \subseteq B$. This verifies (c).

If B is an A^*-group, by hypothesis $O_q(B) = 1$ for all primes $q \in \pi(F(A))$. By (b), it follows that $F(B) \subseteq A$. By 10.6(a), since $C_G(S) \cap F^*(A) \subseteq S \lhd\lhd F^*(A)$, $E(A) \subseteq S \subseteq B$. Part (b) also shows that $[F(B), E(A)] = 1$. Since B is an A^*-group, it follows that $E(A) \subseteq E(B)$. But $|E(A)| \geqq |E(B)|$. Thus $E(A) = E(B) = 1$. Otherwise $A = B$. We may now apply (c) with $\bar{S} = F^*(B) = F(B) \subseteq A$ and get $A = B$.

(e) follows immediately from (d). //

The following two lemmas are easily proved and interesting in themselves. They are also of interest because they seem not to have been noticed in even the solvable case.

Lemma 12.5. _If G is a group such that $F^*(G)$ is a p-group. If U is a p-subgroup of G, then $F^*(C_G(U))$ and $F^*(N_G(U))$ are p-groups also._

Proof. Let $N = N_G(U)$, $C = C_G(U)$. Clearly $F(N)_{p'} \subseteq F(C)_{p'} \subseteq F(N)_{p'}$ and $E(N) \subseteq C$ because $U \subseteq F(N)$. By 10.6 $E(N) = E(C)$. It follows that $O^p(F^*(N)) = O^p(F^*(C))$.

Let $x \in O^p(F^*(N))$ be a p'-element. Then

$$[C_G(U) \cap F^*(G),\ x] \subseteq O^p(F^*(N)) \cap F^*(G) \subseteq Z(F^*(N)).$$

Hence $[C_G(U) \cap F^*(G), x, x] = 1$. By 0.2, $[C_G(U) \cap F^*(G), x] = 1$. Now apply 2.2 to $\langle x \rangle$ acting on $F^*(G)U$. It follows that $[x, F^*(G)] = 1$ and so $x \in F^*(G)$. Hence $x = 1$. Thus $O^p(F^*(N)) = O^p(F^*(C)) = 1$. //

Lemma 12.6. _Let U, V be p-subgroups of a group G such that $V \subseteq U$. Then if $F^*(C_G(V))$ is a p-group, $F^*(C_G(U))$ is a p-group also._

Proof. If $V \trianglelefteq U$, let $N = N_G(V)$. Clearly if $N \subset G$, induction on $|G|$ gives the result since $C_G(U) \subseteq C_G(V) \subseteq N$. Hence we may assume $N = G$.

Now since $F^*(C_G(V))$ is a p-group, $F^*(N_G(V))$ is a p-group also. Apply 12.5 to $N_G(V) = G$ to get the result.

Since $V \lhd\lhd U$, the Lemma follows by induction on the length of a subnormal series from V to U. //

13. PROPERTIES OF A*-GROUPS

Let X be an A*-group, t an involution in X.

Lemma 13. 1. *If a 2-subgroup T of X centralizes O(X), then it is contained in F*(X).*

Proof. Since $[T, O(X)] = 1$, $[T, O(F(X))] = [T, F(O(X))] = 1$. But because the Sylow 2-subgroups of X are abelian, $[T, F(X)] = 1$. But clearly normal subgroups and factor groups of A*-groups are A*-groups.

Hence $F(X)C_X(F(X))/F(X)$ is an A*-group which has no non-trivial solvable normal subgroups. Thus $[F(X)C_X(F(X)) : F^*(X)]$ is odd. Hence $T \subseteq F^*(X)$. //

Lemma 13. 2. *If X is a simple A*-group, then X has one class of involutions.*

Proof. This is well known if $X = L_2(q)$. If X is of type JR, it follows from 12. 1. //

The next two lemmas are concerned with $C_X(t)$-invariant subgroups of X. We are able to locate at least part of them within $F^*(X)$.

Lemma 13. 3. *Let E be a $C_X(t)$-invariant semi-simple subgroup of E(X). Then any component K of E is contained in a component L of E(X). Moreover (i) or (ii) holds.*

(i) $K = L$.

(ii) *K is of type $L_2(q)$, q odd, L is of type JR and t centralizes K.*

Proof. Let T be a Sylow 2-subgroup of X containing t. Since K does not centralize a Sylow 2-subgroup $T \cap E(X)$ of E(X), there exists a component L of E(X) such that $[K, T \cap L] \neq 1$. Now T permutes the components of E(X) and centralizes a Sylow 2-subgroup of each component. Hence T normalizes both K and L. Thus $[K, T \cap L] \trianglelefteq K$ and if $[K, T \cap L] \subseteq Z(K)$, $[K, T \cap L, K] = 1$, whence $[K, T \cap L] = 1$. This is not the case.

It follows that $[K, T \cap L] = K \subseteq L$.

If $K \neq L$, then $C_L(t)$ normalizes $L \cap E \supseteq K$.

If $L \cong L_2(2^n)$, then t must induce an inner automorphism on L since a Sylow 2-subgroup of $P\Gamma L(2, 2^n)$ is non-abelian if n is even. If $L \cong L_2(p^n)$, $p \neq 2$, then n is odd because otherwise a Sylow 2-subgroup of L is dihedral of order $\geqq 8$. If then t were to induce the 'transpose inverse' automorphism on L, a Sylow 2-subgroup of $L\langle t\rangle$ would be non-abelian. Thus t must induce an inner automorphism τ on L if $L \cong L_2(q)$.

Since $C_L(\tau)$ is dihedral if $L \cong L_2(q)$, q odd, and an elementary abelian 2-group if $L \cong L_2(2^n)$, $C_L(\tau)$ normalizes no non-solvable subgroup of L except L itself. Here τ is an involution in L. Hence $K = L$.

If L is of type JR, then t centralizes a Sylow 2-subgroup $T \cap L$ of L and so t normalizes $C_L(\tau)$ for $\tau \in T \cap L$. Now $C_L(\tau) = \langle\tau\rangle \times E$ and so t normalizes $E^{(\infty)} = L_1$, a group of type $L_2(q)$, q odd. As we saw above, t must induce an inner automorphism on L_1 and so there exists an involution $u \in L_1$ such that $[tu, L_1] = 1$. If $t \in L_1 \subseteq L$, then $C_L(t)$ is a maximal subgroup of L and $C_L(t) \supseteq K$. This is the result. Thus $L_1\langle t\rangle = L_1 \times \langle tu\rangle$ and $[tu, E] \subseteq C_L(L_1) \cap C_L(\tau) = \langle\tau\rangle$. It follows that $[tu, E] = 1$ because $[E : L_1]$ is odd.

Now tu centralizes a Sylow 2-subgroup S of L and so tu normalizes $N_L(S)$, which is modulo $C_L(S)$ a non-cyclic group of order 21.

Let $Z/C_L(S)$ have order 7. Let $M = N_L(S)\langle tu\rangle$. Since $tu \in C_M(S) \trianglelefteq M$, $[tu, Z] \subseteq C_M(S) \cap Z \subseteq C_L(S)$. Thus tu centralizes Z modulo $C_L(S)$. It follows that either tu centralizes Z or tu acts on Z like the unique involution $\tau \in S$ which is centralized by an element of order 3 in $N_L(S) \cap C_L(\tau)$. Thus either $[tu, Z] = 1$ or $[tu\tau, Z] = 1$ and so either tu or $tu\tau$ centralizes $N_L(S) \nsubseteq C_L(\tau)$, a maximal subgroup of L by 12.2. Since $u\tau$ is an involution, we lose no generality by assuming that $[tu, L] = 1$ and so $C_L(t) = C_L(u) = \langle u\rangle \times E_1$ for some suitable subgroup E_1. By 12.2, $C_L(u)$ is a maximal subgroup of L. But $C_L(t) = C_L(u)$ normalizes $L \cap E \supseteq K$. Thus $K \subseteq C_L(u) = C_L(t)$ and so K is of type $L_2(q)$. //

Lemma 13.4. Let U be a $C_X(t)$-invariant subgroup of X such that $U = F^*(U)$. Then $[t, U] = V \lhd\lhd F^*(X)$.

Proof. We have already seen that subnormal subgroups V of $F^*(X)$ satisfy $V = F^*(V) = F(V)E(V)$. If we could show that $[t, E(V)] \lhd\lhd F^*(X)$, $[t, F(V)] \lhd\lhd F^*(X)$, then it would follow that $[t, V] = [t, F(V)][t, E(V)] \lhd\lhd F^*(X)$. For $[t, E(V)] \trianglelefteq F^*(X)$ if $[t, E(V)] \lhd\lhd F^*(X)$.

Thus we assume first that $E(U) = U = [t, U]$. We show first that $[U, F(X)] = 1$. Let $D = C_X(U) \cap F(X)$. Then $[C_{F(X)}(t), U] \subseteq U \cap F(X)$, a nilpotent normal subgroup of U and so $[U, C_X(t) \cap F(X)] \subseteq Z(U)$. As usual, $[U, C_X(t) \cap F(X)] = 1$ and so $C_X(t) \cap F(X) \subseteq D$. Thus t inverts $(N_X(D) \cap F(X))/D$. Note that a Sylow 2-subgroup of $F(X)$ lies in $C_X(t) \cap F(X) \subseteq D$.

Now $U = [t, U]$ centralizes $N_X(D) \cap F(X)/D$. Arguing for each Sylow p-subgroup of the nilpotent group $N_X(D) \cap F(X)$ separately, it follows that U centralizes $N_X(D) \cap F(X)$. Thus $D = F(X)$ and $[U, F(X)] = 1$.

Now any Sylow 2-subgroup of U centralizes $F(O(X)) \subseteq F(X)$. Since $C_{O(X)}(F(O(X))) \subseteq F(O(X))$, by 2.2 it follows that any Sylow 2-subgroup of U centralizes $O(X)$. By 12.1 it follows that any Sylow 2-subgroup of U lies in $F^*(X)$ and so $U \subseteq F^*(X)$. Thus $U \subseteq E(X)$. By 13.3(a), $U = [t, U] \lhd\lhd E(X)$. Note that 13.3(b) is not applicable here since if a component K of U is not a component of $E(X)$, then $[K, t] = 1$. But $U = [t, U]$ is just the product of components not centralized by t.

Now assume that $U = [t, U]$ is a p-group where p is an odd prime and that X is a minimal counter example.

First if $O_p(X) \neq 1$, let $\bar{X} = X/O_p(X)$. By induction $[\bar{t}, \bar{U}] \lhd\lhd F^*(\bar{X})$. But any subnormal p-subgroup of $F^*(G)$ is contained in $O_p(G)$ for any group G. Thus $\overline{[t, U]} \subseteq O_p(\bar{X}) = 1$ and so $[t, U] \subseteq O_p(X)$. Thus $O_p(X) = 1$.

Again U centralizes $F(X)$ for otherwise, $C_X(t) \cap F(X) \subseteq C_X(U) \cap F(X) = D$ and U centralizes $N_X(D) \cap F(X)/D$. Thus $[U, F(O(X))] = 1$ and so $[U, O(X)] = 1$ since $(|U|, |F(O(X))|) = 1$. For $[F(O(X)), U, O(X)] = 1$, $[O(X), F(O(X)), U] = 1$ and by the Three Subgroups Lemma, we have $[U, O(X), F(O(X))] = 1$.

Hence $[U, O(X)] \subseteq C_X(F(O(X)) \subseteq F(O(X))$.

Now if $w \in O(X)$, $u \in U$, $w^u = wf$, $f \in F(O(X))$, and so $w^{u^{|u|}} = w = wf^{|U|}$, a contradiction.

If $O(X) \neq 1$, let $X = X/O(X)$. Then by induction $[\bar{t}, \bar{U}] \lhd\lhd F^*(\bar{X})$ and so $[\bar{t}, \bar{U}] \subseteq F(\bar{X})$. Then $[t, U]O(X) \subseteq O_p(X \bmod O(X)) = O(X)$ because $p \neq 2$. Thus $[t, U] = U \subseteq O(X)$. Since $U \subseteq Z(O(X))$ we have a contradiction to $O_p(X) = 1$.

Now X is an A*-group such that $O(X) = 1$. It follows that $t \in F^*(X)$. Hence $U = [t, U] \subseteq F^*(X)$.

If $F^*(X) \subset X$, then $U \lhd\lhd F^*(F^*(X)) = F^*(X)$, by induction. Hence $X = F^*(X)$. If X is not simple, let $N \trianglelefteq X$ be a minimal normal subgroup of X. Since $X = F^*(X)$, N is a simple group. By induction $UN \lhd\lhd X$. But the only subnormal subgroups of X are simple groups or 2-groups (or products of them) and since U is a p-group, $U \subseteq N$. If X is not simple there exists $N_1 \trianglelefteq X$ such that $N_1 \cap N = 1$ and then $U \subseteq N_1 \cap N = 1$. Thus X is a simple A*-group.

Thus either $X \cong L_2(2^n)$ or $C_X(t)$ is a maximal subgroup of X. In the first case, the only odd order group normalized by $C_X(t)$ is 1, while in the second case $U \subseteq C_X(t)$. Since $U = [U, t]$, $U = 1$. This completes the proof. //

Lemma 13.5. Let T be a Sylow 2-subgroup of $C_X(t)$ and $U = OF(C_X(t))$. Then $U \subseteq O^*(X) = F^*(X \bmod O(X))$ and $C_U(T) \subseteq O(X)$. Further for any component E of $E(X)$, the following hold.

(a) U normalizes E;

(b) $U/C_U(E)$ is cyclic;

(c) If E is of type $L_2(2^n)$ or JR, then U centralizes E.

Proof. Induction on $|X|$.

(i) $O(X) = 1$.

Suppose $O(X) \neq 1$. Let $\bar{X} = X/O(X)$, $C = C_X(t)$. Then $C_{\bar{X}}(\bar{t}) = \bar{C}$, $\overline{F(X)} \trianglelefteq F(\bar{X})$ and so $\bar{U} = \overline{OF(C)} \subseteq OF(\bar{C}) = \bar{W}$.

By induction $\bar{U} \subseteq \bar{W} \subseteq O^*(\bar{X}) = \overline{O^*(X)}$ and so $U \subseteq O^*(X)$. Also by induction, $C_{\overline{W}}(\overline{T}) = 1$ and so $C_{\overline{U}}(\overline{T}) = \overline{C_U(T)} = 1$. Hence $C_U(T) \subseteqq O(X)$.

Moreover, if E is a component of $E(X)$, $\overline{E}$ is a component of $E(\overline{X})$. For E normalizes the solvable subgroup $F(X \bmod O(X))$ and so E centralizes $F(\overline{X})$. Since $\overline{E} \lhd\lhd \overline{X}$, it easily follows that $\overline{E} \subseteqq F^*(\overline{X})$, and so $\overline{E}$ is a component of $E(\overline{X})$. Thus $\overline{W}$ normalizes $\overline{E}$ by induction and so W normalizes $EO(X)$. Of course $[E, O(X)] = 1$ because $O(X)$ is solvable and E-invariant and so $E \subseteqq W$ normalizes $(EO(X))^{(\infty)} = E$.

Now put $\overline{D} = C_{\overline{W}}(\overline{E})$. By induction W/D is cyclic and $U/C_U(E) \cong UC_W(E)/C_W(E) \subseteqq W/C_W(E)$. But $[\overline{D}, \overline{E}] = 1$ and so $[D, E] \subseteqq O(X)$. Hence $[E, D, E] = 1$ and $[E, D] = 1$. Thus $D = C_W(E)$. Therefore $U/C_U(E)$ is cyclic.

Again (c) holds since by induction $[\overline{W}, \overline{E}] = 1$ and then $[W, E] = 1$ as usual. Thus $[U, E] = 1$.

(ii) $F^*(X) = O^*(X) \supseteqq T$ where T is a Sylow 2-subgroup of X.

This is clear because X is an A^*-group and $O(X) = 1$.

(iii) U normalizes E and so (a) holds.

For T normalizes each component of $E(X)$ and centralizes a Sylow 2-subgroup $T \cap E(X)$ of $E(X)$. Also $C_U(T)$ permutes the components of $E(X)$ and centralizes $T \cap E(X)$. Thus $C_U(T)$ normalizes E.

But $U = C_U(T)[T, U]$ and by (ii) $T \subseteqq F^*(X)$. Thus $[T, U] \subseteqq F^*(X) \trianglerighteq E$. It follows that U normalizes E.

(iv) $X = F^*(X)U$.

Let $Y = F^*(X)U$ and suppose $Y \subset X$. Then $T \subseteqq C_Y(t) \subseteqq C_X(t)$ and so $U = OF(C_X(t)) \subseteqq Y$. Thus $U \subseteqq OF(C_Y(t))$. By induction $U \subseteqq O^*(Y)$ and $C_U(T) \subseteqq O(Y)$.

Now $[O(Y), F^*(X)] = [O(X), F(X)]$ since $[E(X), O(Y)] = 1$ as usual. Also $[O(Y), F(X)] \subseteqq O(Y) \cap F(X)_2 = 1$ because $F(X)$ is a 2-group. Thus $O(Y) \subseteqq C_X(F^*(X)) \subseteqq F^*(X)$ and $O(Y) \trianglelefteq F^*(X)$. This is absurd. Hence $O(Y) = 1$, $U \subseteqq F^*(Y) = F^*(X)$, $C_U(T) = 1$. Parts (b), (c) clearly hold in Y and so in X. Thus $Y = X$.

(v) $X = EU$, where E is a non-abelian simple group.

First if $F^*(X) = F(X)$ is a 2-group, then $X = F(X)U = F(X) \times U$

since $[U, F(X)] \subseteq F(X) \cap U = 1$. This contradicts $O(X) = 1$.

Let $F^*(X) = EE_1$ where E is a component of $E(X)$, and let $Y = EUT$. Note U normalizes E by (a). First $O(Y) \cap E = 1$ and so $[t, O(Y)] = 1$ because $[t, U] = 1$. Since Y/ET is nilpotent, $O(Y)$ is nilpotent. Finally $C_Y(t) = UT\,C_E(t) \subseteq C_X(t)$. Thus $U \subseteq OF(C_Y(t))$. Since $[E, O(Y)] = 1$, $EO(Y) = E \times O(Y)$.

By induction if $Y \subset X$, $U \subseteq O^*(Y)$, $C_U(T) = 1$ and (b), (c) hold. $O^*(Y) = ETO(Y)$ clearly and so $U \subseteq EC_U(E)$.

Arguing similarly on $Y_1 = E_1UT$ we have $U \subseteq E_1C_U(E_1)$. Thus $U \subseteq E_1C_U(E_1) \cap E\,C_U(E) = EE_1C_U(EE_1)$. Hence $U \subseteq F^*(X)$ and the theorem is true.

We may assume that $X = EUT$. Let τ be the projection of t on E. If $\tau = 1$, then $[t, E] = 1$ and so $t \in Z(X)$. But $O(X) = 1$ and $U \subseteq O(C_X(t))$. Thus $\tau \neq 1$. Since $[U, \tau] = 1$, $C_X(t) = C_X(\tau)$. If $Y = EU \subset X$, we may induct to Y, τ in place of X, t and get $U \subseteq O^*(Y) = E \subseteq F^*(X)$ and $C_U(T \cap E) = 1$. Since $C_U(T \cap E) = C_U(T)$, we have (v).

(vi) E is of type $L_2(q)$, q odd.

If E is of type $L_2(2^n)$, then $C_E(t) = T$ and $[U, C_E(t)] \subseteq U \cap E$ and so $[U, C_E(t)] \subseteq O(C_E(t)) = 1$.

Let $\bar{U} = U/C_U(E) \subseteq P\Gamma L(2, 2^n)$ and $[\bar{U}, T] = 1$. At least one of the elements of T is moved by any field automorphism and so $\bar{U}$ must induce inner automorphisms on E. Since $C_E(T) = T$, $\bar{U} = 1$ and $U = C_U(E)$. This contradicts $O(X) = 1$ since $EU = X = E \times U$.

If E is of type JR, then $C_E(t) = \langle t \rangle \times H$ where $L_2(q) \subseteq H \subseteq P\Gamma L(2, q)$. Therefore $O(C_E(t)) = 1$. Since $[U, C_E(t)] \trianglelefteq C_E(t) \cap U$ we must have $[U, C_E(t)] = [U, T] = 1$. Let $s \in T^{\#}$, $s \neq t$. Note $U \subseteq C_X(s)$ and so $C_X(s) = C_E(s)U = (\langle s \rangle \times L)U$ where L is a conjugate of H. Clearly $C_X(s)$ is an A^*-group and $C_X(s) \subset X$.

Since $[T, U] = 1$, $U = C_U(T)$. Now $U \subseteq OF(C_X(t) \cap C_X(s))$ and by induction $U = C_U(T) \subseteq O(C_X(s))$. Then $[U, C_E(s)] \subseteq O(C_E(s)) = 1$. Thus U centralizes $\langle C_E(t), C_E(s) \rangle = E$ by 12.2. Again this contradicts $O(X) = 1$. This completes the proof of (vi).

(vii) The final contradiction.

By the above results $X = EU$, $E \cong L_2(q)$, q odd. Therefore $C_E(t)$ is dihedral of order $q - \varepsilon$, where $q \equiv$ (mod 4), $\varepsilon = \pm 1$, T is of type (2, 2).

If $U \subseteqq E$ we are done because $E = O^*(X)$, $C_U(T) = U \cap C_E(T) = T \cap U = 1$ and $U = O(C_E(t))$ is cyclic.

As $U = C_U(T)[U, T]$ and $[U, T] \subseteqq E$, it follows that $C_U(T) \neq 1$. Let $u \in C_U(T)$ have prime order p. Let $q = r^n$, where r is a prime. First u must induce a field automorphism on E, since $C_E(T) = T$. Thus $C_E(u) \cong L_2(r^m)$ where $m = n/p$.

Let $C_E(t) = D\langle x\rangle$ where D is cyclic and x inverts D. Then $[u, O_{p'}(D)] \subseteqq O_p(C_X(t)) \cap O_{p'}(D) = 1$. Thus u centralizes $O_{p'}(D)$, T. On the other hand, u normalizes $O_p(D)$, a cyclic p-group. Thus $p|C(u) \cap O_p(D)| \geqq |O_p(D)|$. Thus $[C_E(t) : C_E(t) \cap C_E(u)] \leqq p$. Now $|C_E(t) \cap C_E(u)| = r^m \pm 1$ and so $r^{mp} \pm 1 \leqq p(r^m \pm 1)$. This inequality is not solvable with $r \geqq 3$, $p \geqq 3$. Lemma 13. 5 is completely proved. //

Lemma 13. 6. <u>If</u> U <u>is an abelian subgroup of</u> X <u>of type</u> (2, 2) <u>and</u> E <u>is a component of</u> $E(X)$ <u>not of type</u> $L_2(2^n)$, <u>then</u> $E = \langle C_E(u) : u \in U^\# \rangle$.

Proof. Each $u \in U^\#$ normalizes E by 13. 5 and induces an inner automorphism on E. Further if $u_1, u_2 \in U^\#$, $u_1 \neq u_2$, then u_1, u_2 induce different inner automorphisms on E. But if E is not of type $L_2(2^n)$, then the centralizers of any two distinct involutions of E are distinct maximal subgroups of E. Hence the result. //

14. PROOF OF THE THEOREM A, PART I

Let G be henceforth a minimal counter example to Theorem A. We show as an initial reduction that G is a non-abelian simple group all of whose proper subgroups are A*-groups.

Thus let N be a minimal normal subgroup of G. Clearly $O(G)=1$ and $O^{2'}(G) = G$ because otherwise G is immediately an A*-group. Thus $|N|$ is even.

If N is a 2-group, then $G = C_G(N)$ because $G/C_G(N)$ has odd order. Thus $|N| = 2$. Now G/N is an A*-group by induction. Since $O(G/N) = 1$ and $O^{2'}(G) = G$, it follows that $G/N = S/N \times L_1/N \times \ldots \times L_k/N$, where S/N is a 2-group and L_i/N are simple A*-groups for $i = 1, 2, \ldots, k$.

Because the Sylow 2-subgroups of G are abelian, it follows by an elementary transfer argument, see [12], that $G' \cap N = 1$ and so $G' \cap S = 1$. Clearly G' is a perfect A*-group with $O(G') = 1$ and so G' is a direct product of simple A*-groups. Since $S \trianglelefteq G$, $G = G' \times S$ and G is an A*-group, a contradiction.

Thus we may assume that N is a direct product of isomorphic simple A*-groups L_i, $N = L_1 \times \ldots \times L_k$.

Clearly $C_G(N) \trianglelefteq G$, $C_G(N) \cap N = 1$ and so $C_G(N)$ is an A*-group which has no non-trivial solvable normal subgroups. Thus $NC_G(N) = N \times C_G(N)$ and $G/NC_G(N)$ acts as a group of automorphisms of N. Since a Sylow 2-subgroup T of G is abelian and T permutes the components $L_1, \ldots, L_k$ of N centralizing $T \cap L_i$, a Sylow 2-subgroup of L_i for all $i = 1, \ldots, k$, it follows that T normalizes each component.

We show that a simple A*-group has no non-trivial outer automorphism x of order 2 which centralizes a Sylow 2-subgroup.

For simplicity write $L = L_1$ and suppose that L is of type $L_2(p^n)$. Clearly x must induce a field automorphism on L since a Sylow 2-subgroup of $PGL(2, q)$ is dihedral and non-abelian if q is odd, and so $n = 2m$ is even. But then a Sylow 2-subgroup of $L_2(p^{2m})$ is non-abelian.

Thus we may assume that L is of type JR. Let S be a Sylow 2-subgroup of L, x an involution which induces an automorphism on L trivial on S. Let $t \in S$ be an involution and suppose $C_L(t) = \langle t \rangle \times E$ where $F = L_2(q) \subseteq E \subseteq P\Gamma L(2, q)$ and q is odd. As above it follows that x must induce an inner automorphism on F. Hence there exists $f \in F$ such that $y = xf$ centralizes F and t.

It is easy to see that y must act trivially on E. For $[y, E] \subseteq C_L(F) \cap C_L(t) = \langle t \rangle$ and so $[y, E] = 1$ because $[E : F]$ is odd. Thus y centralizes $C_L(t)$.

Now y centralizes S and so normalizes $N_L(S)$. By Burnside's Transfer Theorem, $N_L(S) \subseteqq C_L(t)$. It follows that $N_L(S)/C_L(S)$ is an odd order subgroup of $GL(3, 2)$ of order > 3. Hence $N_L(S)/C_L(S)$ is a non-cyclic group of order 21.

Let $Z/C_L(S) \trianglelefteq N_L(S)/C_L(S)$ have order 7. Either y centralizes Z or y acts on $N_L(S)$ like the unique involution in S which is centralized by an element of order 3 in $N_L(S) \cap C_L(t)$, viz. t. Thus either $[y, N_L(S)] = 1$ or $[yt, N_L(S)] = 1$. In both cases, since $[yt, C_L(t)] = 1$, we have $L\langle x\rangle = L \times \langle z\rangle$, where $z^2 = 1$, because $C_L(t)$ is a maximal subgroup of L by 12.2.

This shows that $TL_i \subseteqq L_i C_G(L_i)$ for all i and so $TN \subseteqq NC_G(N)$. Thus $G = NC_G(N) = N \times C_G(N)$ because $O^{2'}(G) = G$, and then G is an A*-group if G is not simple.

Thus G is a simple group all of whose proper subgroups are A*-groups. We remark that a Sylow 2-subgroup T of G has rank at least 3. For otherwise $r(T) = 2$ and $|T| = 4$ by [7]. Then $G \cong L_2(q)$ by [13]

The proof of Theorem A proceeds by showing that either G is itself an A*-group or G has a strongly embedded subgroup. But in that case, by [4], $G \cong L_2(2^n)$, $Sz(2^{2n+1})$ or $U_3(2^n)$. Of these only the groups $L_2(2^n)$ have abelian Sylow 2-subgroups and so G is an A*-group every time. The proof of Theorem A will then be completed.

We study the structure of the minimal counter example G to Theorem A by considering a collection of maximal subgroups containing $C_G(t)$, where t is an involution of G. The maximal subgroups studied are ingeniously chosen in order to enable the exploitation of the uniqueness Theorem B of Chapter 12 and its immediate consequence Theorem 12.4. As Bender points out in [5], the explicit definition of the class of maximal subgroups used so frequently in the proof is required for only one result, 14.1. In that result 14.1, we identify a collection of subnormal subgroups of $F^*(H)$, where H is a certain maximal subgroup containing $C_G(t)$. These subnormal subgroups are ones over whose G normalizers we have some control. They assume crucial importance in the results which follow.

Thus let G be a simple group with abelian Sylow 2-subgroups all of whose proper subgroups are A*-groups. Assume that G contains a 3-generated abelian 2-subgroup and that G does not contain a strongly embedded subgroup.

Let t be an involution in G. Let H be a maximal subgroup of G containing $C_G(t)$. Choose H in such a way that for some prime p, $O_p(H) \neq 1$ but $C_G(t) \cap O_p(H) = 1$. If, for no prime p, can we find a maximal subgroup $H \supseteq C_G(t)$ with a subgroup $O_p(H)$ inverted by t, choose H such that $|E(H)|$ is maximal. Let M(t) be the set of all maximal subgroups of G containing $C_G(t)$ satisfying the above conditions. Note that if for some $H \in M(t)$, $O_p(H) \neq 1$ and $C_G(t) \cap O_p(H) = 1$, then every subgroup $M \in M(t)$ has for some (possibly different) prime q, $O_q(M) \neq 1$ and $C_G(t) \cap O_q(M) = 1$.

Let T be a fixed Sylow 2-subgroup of G containing t and let $\pi = \pi(F(H))$.

We now define a set $\mathfrak{A} = \mathfrak{A}(H)$ of subnormal subgroups of $F^*(H)$ satisfying a kind of uniqueness condition. Let $U \lhd\lhd F^*(H)$. If $|\pi(F^*(H))| \geqq 2$, then $U \in \mathfrak{A}$ if and only if $N_G(U) \subseteq H$. If $|\pi(F^*(H))| = 1$, then $U \in \mathfrak{A}$ if and only if $\pi(F^*(C_G(U)) = \pi(F^*(H))$. Note that if $U \lhd\lhd F^*(H)$ and $N_G(U) \subseteq H$, then $U \in \mathfrak{A}$ every time. For if $F^*(H)$ is a p-group and $N_G(U) = N_H(U)$, then by 12.7, $F^*(C_G(U)) = F^*(C_H(U))$ is a p-group also.

Here in 14.1, we get criteria which allow us to recognize elements of $\mathfrak{A}(H)$ more easily.

Lemma 14.1. Let $V \neq 1$ be a t-invariant subnormal subgroup of $F^*(H)$. Each of the following conditions implies that $V \in \mathfrak{A}(H)$.

(a) V is $C_G(t)$-invariant.

(b) Some subgroup $U \in \mathfrak{A}$ centralizes V and satisfies $[U, t] = U$.

(c) There is a non-cyclic abelian subgroup U of odd order, where $U \subseteq C_G(t) \cap C_G(V)$ such that $\langle u \rangle \in \mathfrak{A}$ for all $u \in U^{\#}$.

Proof. Assume that $B \subset G$, $B \supseteq N_G(V)$.

Since $E(H) \subseteq S = N_G(V) \cap F^*(H)$ for all $X \lhd\lhd F^*(H)$, $S \lhd\lhd F^*(H)$. Also $C_G(S) \cap F^*(H) \subseteq C_G(V) \cap F^*(H) \subseteq S$. Apply 12.4(a) and get

$O_q(B) \cap H = 1$ for all $q \in \pi(F(H))'$. Thus since $C_G(t) \subseteq H$ and t normalizes $F(B)_{\pi'}$, t must invert $F(B)_{\pi'}$.

Consider case (b). Let $W = B \cap F^*(H) \supseteq N_G(V) \cap F^*(H) \supseteq E(H)$. Now we may apply 13.4 with $X = B$ and get $[t, W] \lhd\lhd F^*(B)$. Now $U \lhd\lhd W$ and so $U = [t, U] \lhd\lhd [t, W]$. Hence $U \lhd\lhd F^*(B)$.

If $|\pi(F^*(H))| = 1$, put $B = N_G(V)$, $\{p\} = \pi(F^*(H))$. Then $E(B) \subseteq C_G(V) \subseteq C_G(U)$, since $U \lhd\lhd F^*(B)$ and is a p-group. (Remember any subnormal p-subgroup of F^* lies in F.) Now $F(C_G(U) \cap C_G(V))$ is a solvable $E(B)$-invariant subgroup of B and so $[E(B), F(C_G(U) \cap C_G(V))] = 1$. Since $C_G(U) \cap C_G(V)$ is an A^*-group, $E(B) \subseteq E(C_G(U) \cap C_G(V))$.

Now $[F(B)_{p'}, V] \subseteq V \cap F(B)_{p'} = 1$, since $V \subseteq F^*(H)$ is a p-group. Also $[F(B)_{p'}, U] \subseteq F(B)_{p'} \cap F(B)_p = 1$, since U is a subnormal p-subgroup of $F^*(B)$.

Thus $O^p(F^*(B)) \subseteq O^p(F^*(C_G(U) \cap C_G(V))$.

But by 12.5, if $F^*(C_G(U))$ is a p-group, then $F^*(C_G(U) \cap C_G(V)$ is a p-group also. Thus $F^*(B)$ is a p-group also. By 12.5 again, $F^*(C_G(V))$ is a p-group and so $V \in \mathfrak{A}$.

If $|\pi(F^*(H))| \geq 2$, let $B \supseteq N_G(V)$ be a maximal subgroup of G. We know $U \lhd\lhd F^*(B)$ and so $N_G(U) \supseteq E(B)$. Thus $H \cap F^*(B) \supseteq E(B)$ because $N_G(U) \subseteq H$ in this case. It follows that $U_1 = H \cap F^*(B) \lhd\lhd F^*(B)$.

Now $C_G(U_1) \cap F^*(B) \subseteq C_G(U) \cap F^*(B) \subseteq H \cap F^*(B) = U_1$.

Taking $S = N_G(V) \cap F^*(H) \supseteq E(H)$, we have first $S \lhd\lhd F^*(H)$, then $S \subseteq B$, $C_G(S) \cap F^*(H) \subseteq S$. Apply directly 12.4(c) and get $H = B \supseteq N_G(V)$ and $V \in \mathfrak{A}(H)$. This verifies (b).

Suppose now (b) does not apply. Let B be a maximal subgroup of G containing $N_G(V)$. By 12.4(a), $O_q(B) \cap H = 1$ for all $q \in \pi(F(H))'$. Thus since $C_G(t) \subseteq H$ and t normalizes $O_q(B)$, if $F(B)_{\pi'} \neq 1$, t must invert some non-trivial $O_q(B)$ for some q. But then t must have inverted some non-trivial $O_r(H)$ for some prime r! But then $U = O_r(H)$ is abelian and contained in $Z(F^*(H))$. Clearly $U = [U, t]$ and $N_G(U) = H$. Hence $U \in \mathfrak{A}$. Since $[V, u] = 1$, (b) shows that $V \in \mathfrak{A}(H)$. Thus we may assume that $F(B)_{\pi'} = 1$.

Moreover, t does not invert any subgroup $O_p(H)$ for any prime p. For then $O_p(H) \in \mathfrak{A}(H)$, $O_p(H) \subseteq Z(F^*(H))$ and by (b) $V \in \mathfrak{A}(H)$. Since

$H \in M(t)$, $E(H)$ must then be of maximal possible order. But B is a maximal subgroup containing $C_G(t)$ if (a) applies. Since $H \in M(t)$, $|E(H)| \geqq |E(B)|$. It is clear that we have directly reproduced the hypotheses of 12.4(d) since H, B are A*-groups. Thus we have $H = B \supseteqq N_G(V)$ and $V \in \mathfrak{A}(H)$ if $|\pi(F^*(B))| \geqq 2$ since otherwise 12.4 does not apply. However, if $\pi(F^*(B)) = \{p\}$, then $F^*(C_G(V))$ is a p-group also by 12.5. Thus in any case $V \in \mathfrak{A}(H)$.

In case (c), $\langle u \rangle \lhd\lhd F^*(H)$, $u \in U^{\#}$. Since $\langle u \rangle$ is cyclic, $\langle u \rangle \subseteqq F(H)$ for all $u \in U^{\#}$. Thus $U \subseteqq F(H)$.

Suppose $|\pi(F^*(H))| \geqq 2$. Let B be a maximal subgroup of G containing $N_G(V)$. Since $U \subseteqq OF(C_B(t))$. Apply 13.5(b) and get that every component of $E(B)$ is centralized by some non-trivial element of U. Thus if $N = O^{\pi}(F^*(B))$, then $N = \langle C_N(u) : u \in U^{\#} \rangle$. Since $\langle u \rangle \in \mathfrak{A}(H)$ and $|\pi(F^*(H))| \geqq 2$, $C_G(u) \subseteqq H$. Thus $N \subseteqq H$. By 12.4(b), $F(B)_{\pi} \subseteqq H$. Hence $F^*(B) \subseteqq H$. We may apply 12.4(c) since $N(V) \cap F^*(H) \subseteqq B$ to get $V \in \mathfrak{A}(H)$.

If $F^*(H)$ is a p-group and $V \notin \mathfrak{A}(H)$, then $F^*(C_G(V))$ is not a p-group. As remarked above, every component of $E(B)$, where $B = C_G(V)$, is centralized by a non-trivial element $u \in U$. Since $\langle u \rangle \in \mathfrak{A}(H)$, $F^*(C_G(u))$ is a p-group. But $V \subseteqq F^*(H)$ is a p-group and by 12.5, $F^*(C_G(\langle u \rangle V)) = F^*(C_B(u))$ is a p-group. But clearly $C_G(u) \cap O^p(F^*(C_G(V)))$ contains non-trivial p'-elements for suitable $u \in U^{\#}$. Since $C_G(u) \cap O^p(F^*(C_G(V)) \subseteqq F^*(C_B(u))$ by 10.7, we have the required contradiction. This completes the proof of 14.1. //

Lemma 14.2. H <u>has at least</u> 2 <u>classes of involutions.</u>

Proof. If $g \in G$, $t \in H \cap H^g$, where exists $h \in H$ such that $t^{g^{-1}h} = t$, if H has one class of involutions. But then $g^{-1}h \in C_G(t) \subseteqq H$ and so $g \in H$. Thus, if H has one class of involutions, $H \cap H^g$ has odd order, if $g \in G - H$. Hence H is strongly embedded in G, a contradiction. //

Lemma 14.3. <u>Let</u> R <u>be a subgroup of</u> T <u>such that</u> $r(R) = r(T) - 1$. <u>Then</u> $M(s) \neq \{H\}$ <u>for some involution</u> $s \in R$.

Proof. Suppose $M(s) = \{H\}$ for all $s \in \Omega_1(R)^{\#}$. First if $M(s) \Rightarrow M(s^g) = M(s)^g = \{H\}$, then clearly $g \in N_G(H) = H$.

Let $n \in N_G(T)$. Since $r(T) \geqq 3$, $R \cap R^n \neq 1$ and so there exists an involution $s \in R \cap R^n$. Then $s, s^{n^{-1}} \in R$ and by assumption $M(s) = M(s^{n^{-1}})$. Thus $n \in H$, $N_G(T) \subseteqq H$.

But every involution of G is conjugate by an element of $N_G(T)$ to an element of R. Hence $M(s) = \{H\}$ for all involutions $s \in H$. Now if $s \in H \cap H^g$, where s is an involution, then $g \in N_G(H) = H$ and H is strongly embedded in G. //

Lemma 14.4. Let D be a T-invariant subgroup of odd order such that $[t, D] \neq 1$. Then T contains a subgroup R such that $r(R) = r(T) - 1$ and $[t, C_D(R)] \neq 1$.

Proof. Without loss of generality, D is a p-group and $C_D(t) \trianglelefteq D$ and T acts irreducibly and non-trivially on $D/C_D(t)$. Since T is abelian, T induces a cyclic group of automorphisms on $D/C_D(t)$. Also t acts non-trivially on $D/C_D(t)$. Let $R = C_T(D/C_D(t))$. Then clearly $r(R) = r(T) - 1$. Also $C_D(R).C_D(t) = D$ since $C_{D/C_D(t)}(R) = C_D(R)C_D(t)/C_D(t)$. Since $[t, D] \neq 1$, $[r, C_D(R)] \neq 1$. //

Lemma 14.5. Let p be an odd prime such that $[t, O_p(H)] \neq 1$. If $P = O_p(H)$, then P has a $C_P(t)$ invariant subgroup $\overline{P}$ such that $[t, \overline{P}] \neq 1$ and $V \in \mathfrak{A}(H)$ for any t-invariant subgroup V of $\overline{P}$, $V \neq 1$.

Proof. Suppose the result is false.

Let $V = C_P(C_P(t))$.

First if $[t, V] = 1$, then $[t, P] = 1$ by 2.2. This is not the case. Thus $[t, V]$ is a $C_P(t)$-invariant subnormal subgroup of $F^*(H)$. Thus $U = [t, V] \in \mathfrak{A}(H)$ by 14.1(a). Now for any subgroup W of $C_P(t)$ we have $[W, U] = 1$. Since $U \in \mathfrak{A}(H)$, $W \in \mathfrak{A}(H)$ by 14.1(b).

If $C_P(t)$ is non-cyclic, there exists a non-cyclic normal 2'-subgroup U_1 of type (p, p) of $C_P(t)$, which is, as seen above, an element of $\mathfrak{A}(H)$. So by 14.1(c), since $U_1 \subseteqq C_P(t) \cap C_P(V)$, $V \in \mathfrak{A}(H)$. Finally if Y is any t-invariant subgroup of V, $Y \in \mathfrak{A}(H)$ by the same argument.

Letting $\overline{P} = V$ we have the result.

If $C_P(t)$ is cyclic, first every $C_P(t)$-invariant abelian subgroup A of P is centralized by t. For otherwise, $[t, A] \in \mathfrak{a}(H)$ by 14. 1(a) and then any t-invariant subgroup of $[t, A]$, being centralized by $[t, A] \in \mathfrak{a}(H)$, is itself an element of $\mathfrak{a}(H)$ by 14. 1(b).

Therefore every $C_P(t)$-invariant abelian subgroup of P is contained in $C_P(t)$ and so is cyclic. It follows that $Z(P')$ is cyclic. Now any normal subgroup Q of P of type (p, p) lying in P' is clearly contained in $Z(P')$. For $P/C_P(Q) \subseteq GL(2, p)$ and so has order $\leq p$. Thus $P' \subseteq C_P(Q)$. It follows that P' is cyclic since p is odd. Further $P' \subseteq C_P(t)$.

Now if $x \in P$, $x^t = x^{-1}$, then for $y \in P'$ we have $y^{xt} = y^{x^{-1}} = y^x$. Thus $[x^2, y] = 1$ and $[x, y] = 1$. Since $P = C_P(t)I$ where $I = \{x \in P : x^t = x^{-1}\}$, $P' \subseteq Z(P)$. Let $Q = \Omega_1(P)$. Since P has class ≤ 2, Q has exponent p and $Z(Q)$, being cyclic, has order p. Also $C_Q(t) = Z(Q)$.

Every non-abelian subgroup M of Q is an element of $\mathfrak{a}(H)$. For $M' = Z(Q)$ and so $N_G(M) \subseteq N_G(M') = H$. On the other hand, if $L \neq 1$ is an abelian t-invariant subgroup then either $L \subseteq C_Q(t) = Z(Q)$ and $L \in \mathfrak{a}(H)$, or $[t, L] \neq 1$. If $x \in L$ is such that $x^t = x^{-1}$, then $L \in \mathfrak{a}(H)$ if $\langle x\rangle \in \mathfrak{a}(H)$ by 14. 1(b). Thus either every t-invariant subgroup of Q lies in $\mathfrak{a}(H)$ or there exists $x \in Q$ such that $x^t = x^{-1}$, $\langle x\rangle \notin \mathfrak{a}(H)$.

Consider $[t, C_Q(x)] = V$. If $V \in \mathfrak{a}(H)$ then so is $\langle x\rangle$ by 14. 1(b), a contradiction. Hence V is abelian. But

$$C_Q(x) = (C_Q(t) \cap C_Q(x))[t, C_Q(x)] \subseteq Z(Q)[t, C_Q(x)] = Z(Q)V.$$

Thus $C_Q(x)$ is abelian and of index p in Q. For the centralizer of any non-central element of Q is of index p in Q. Since $C_Q(x)$ is not cyclic, it is not $C_P(t)$-invariant. Thus Q has two abelian maximal subgroups and so $|Q| \leq p^3$.

Now let $R \subseteq T$, $r(R) = r(T) - 1$ and $C_Q(R) \not\subseteq Z(Q)$. Such an R exists since we could take $R \subseteq C_T(Q_1/\Phi(Q))$ where Q_1 is an irreducible T submodule of $Q/\Phi(Q)$.

Choose $g \in G$ such that $t^g \in R$. Then $C_G(t^g) \nsubseteq Z(Q)$ and so $H^g \neq H$. Put $K = [t, C_Q(t^g)] \subseteq H \cap H^g$ since $C(t^g) \subseteq H^g$. Let K_1 be the normal closure of K under $C_G(t) \cap H^g$. Since $K \subseteq Q$, $K_1 \subseteq Q$ and is a p-group. By 12.4, $[t, K_1] \lhd\lhd F^*(H^g)$ and $K = [K, t] \subseteq [K_1, t]$. Since $[K_1, t]$ is a p-group, $K \lhd\lhd F^*(H^g)$. Thus $K \subseteq F(H^g)$ and so $K \subseteq Q^g$.

If $K = Q^g$, then $K = Q$ since $K \subseteq Q$. Then $g \in N(Q) = H$, a contradiction since $H^g \neq H$. Thus $K \subset Q^g$, $|K| \leq p^2$. But $K = [K, t]$ and so $C_K(t) = 1$. Thus $|K| = p$. But $K \subseteq C_G(t^g)$ and so $K = Z(Q^g)$. Therefore $N_G(K) \subseteq H^g$.

Now H, H^g are conjugate A^*-groups and $O^p(F^*(H)) \subseteq C_G(K) \subseteq H^g$. Thus if $S = N_G(K) \cap F^*(H)$, $S \supseteq E(H)$ and $S \lhd\lhd F^*(H)$. Also $C_G(S) \cap F^*(H) \subseteq S$.

If $|\pi(F^*(H))| \neq 1$, then $H = H^g$ by 12.4(e), a contradiction. Therefore $F^*(H) = P$ and T acts faithfully on P and also on Q and also on $Q/\Phi(Q)$. Since then $T \subseteq GL(2, p)$ and $r(T) \geq 3$, we have a contradiction. //

Lemma 14.6. *If $F^*(H)$ is a p-group, where p is an odd prime, then for any involution $s \in G$ and any $M \in \mathcal{M}(s)$, $F^*(M)$ is also a p-group.*

Proof. From 14.5, there is a $C_{O_p(H)}(t)$-invariant subgroup $\overline{P}$ of $O_p(H)$ such that $[t, \overline{P}] \neq 1$ and $V \in \mathfrak{A}(H)$ for every t-invariant subgroup V of $\overline{P}$. Now by 14.4, there exists a subgroup $R \subseteq T$ such that $r(R) = r(T) - 1$ and $W \in [t, C_{\overline{P}}(R)] \neq 1$. Of course $W \in \mathfrak{A}(H)$. By 12.1, we may assume that our involutions $s \in R$. Let $M \in \mathcal{M}(s)$. Let $W_1 = W^{C_M(t)}$, the subgroup generated by the $C_M(t)$-conjugates of W. Since $W \subseteq O_p(H)$, $C_G(t) \subseteq H$, $W_1 \subseteq O_p(H)$ and so W_1 is a p-group. By 12.4, $[t, W_1] \lhd\lhd F^*(M)$. Thus $W = [t, W] \subseteq W_1$ is a subnormal subgroup of $F^*(M)$. Hence $W \subseteq O_p(M)$.

If $P = O_p(M)$, $U = N_P(W)$, since $F^*(N_G(W))$ is a p-group ($W \in \mathfrak{A}(H)$!), $F^*(C_G(U))$ is a p-group also by 12.5. For $C_G(U) \subseteq C_G(W) \subseteq N_G(W)$. Now by 12.6, $F^*(C_G(P))$ is a p-group also since $U \subseteq P$. It follows that $F^*(N_G(P)) = F^*(M)$ is a p-group also.

This completes the proof. //

The following major Theorem is the cornerstone on which the whole proof rests. Its proof is at least staggering in its power and originality. Of course, it is due entirely to Bender.

Theorem 14.7. *Let* p *be an odd prime such that* $[t, O_p(H)] = 1$. *Let* V *be an elementary abelian* p-*subgroup of* $O_p(H)$ *such that* $r(V) \geqq 3$. *Then* $C_G(v) \subseteqq H$ *for all* $v \in V^{\#}$.

Proof. If $[t, F^*(H)] \neq 1$, then either $W_1 = [t, O_q(H)] \neq 1$ for some prime $q \neq p$, or $W_2 = [t, E(H)] \neq 1$ is an element of $\mathfrak{A}(H)$ by 14.1(a). For both W_1, W_2 are $C_H(t)$-invariant and subnormal in $F^*(H)$. But then $\langle v \rangle \subseteqq V \subseteqq O_p(H)$ centralizes W_1 or W_2 and then 14.1(b) implies that $\langle v \rangle \in \mathfrak{A}(H)$ for $v \in V - 1$. Then $C_G(v) \subseteqq N_G(\langle v \rangle) \subseteqq H$ since $F^*(H)$ is not a p-group.

Thus $t \in C_H(F^*(H)) = Z(F^*(H))$. Let $W = C_H(V) \cap F^*(H)$.

Let $\mathfrak{U}(X)$, where $X \subseteqq G$, be the set of all W-invariant A^*-subgroups K of X such that $K = F(K)_{\pi'}E(K)$ and no component of $E(K)$ is contained in H.

If $\mathfrak{U}(G) = \{1\}$, then for any maximal subgroup B of G, $B \supseteqq W$, it follows that $F(B)$ is a π-group and $E(B)$ is a product of components which all lie in H. For if a component of $E(B)$ does not lie in H, then neither does any of its W-conjugates. Then $E^W \in \mathfrak{U}(G) \neq \{1\}$.

Thus $E(B) \subseteqq H$ and $F(B) \subseteqq H$ by 12.4(b). Note that $O^p(F^*(H)) \neq 1$ since $t \in F^*(H)$. Thus $F^*(B) \subseteqq H$. But $C_G(W) \cap F^*(H) \subseteqq W$ and $W \subseteqq B$. By 12.4(c), we have $B = H$ and H is the only maximal subgroup of G containing W. Clearly if $v \in V^{\#}$, $C_G(v) \supseteqq W$ and so $C_G(v) \subseteqq H$. Thus we may assume that $\mathfrak{U}(G) \neq \{1\}$.

We derive a contradiction to the simplicity of G from $\mathfrak{U}(G) \neq \{1\}$ by a series of steps.

(i) If $W \subseteqq X \subset G$, then $[t, F^*(X)] \in \mathfrak{U}(X)$.

For if $W \subseteqq X \subseteqq G$, then t inverts $F(X)_{\pi'}$ elementwise by 12.4(a) since $t \in W$. Also t centralizes $F(X)_{\pi'} \subseteqq H$ by 12.4(b) since if $q \in \pi$, $[O_q(X), O^q(F^*(H))] = 1$ and $t \in O^q(F^*(H)) \neq 1$. Now $[t, E(X)]$ is the product of those components of $E(X)$ not centralized by t, and since

$H/C_H(O_2(H))$ has odd order, $[t, E(X)]$ is the product of those components of $E(X)$ which do not lie in H. For any component of $E(X)$, which lies in H, lies in $C_G(O_2(H)) \subseteq C_G(t)$. Note $t \in Z(F^*(H))$.

Consider now $[t, F^*(X)]$. This group is a W-invariant normal subgroup of $F^*(X)$ and by the argument of the previous paragraph and the structure of normal subgroups of F^*, $[t, F^*(X)] \in \mathcal{U}(X)$.

(ii) $\langle \mathcal{U}(X) \rangle \in \mathcal{U}(X)$ if $X \subset G$.

Let $K \in \mathcal{U}(X)$ where $X \subset G$. Then K centralizes $F(X)$. For $[K, F(X)_\pi, t] \subseteq [F(X)_\pi, t] = 1$ since t centralizes $F(X)_\pi$ by 12.4(b).

$[F(X)_\pi, t, K] = 1$.

Hence $[K, t, F(X)_\pi] = 1$ and $K = [K, t]$ centralizes $F(X)_\pi$.

Also t inverts $F(X)_{\pi'}$, and so commutes with K in its action on $F(X)_{\pi'}$. Thus $K = [K, t]$ centralizes $F(X)_{\pi'}$. Hence $[K, F(X)] = 1$.

Now if $X \subset G$, X is an A^*-group, $t \in O^*(X)$ and so $K = [K, t] \subseteq O^*(X) \cap C_G(F(X))$. It follows that $K \subseteq F^*(X)$.

Thus $K = [K, t] \subseteq [F^*(X), t] \in \mathcal{U}(X)$ by (i). Hence $\langle \mathcal{U}(X) \rangle \subseteq [F^*(X), t] \in \mathcal{U}(X)$.

(iii) There exists $R \subseteq V$ such that V/R is cyclic and $\mathcal{U}(C_G(R)) \neq \{1\}$.

First replacing V by $V\Omega_1(Z(O_p(H)))$ if necessary we may assume that $C_G(V) \subseteq H$. Choose $K \in \mathcal{U}(G)$. If $F(K) \neq 1$, then $[V, F(K)] \neq 1$. For otherwise $F(K) \subseteq C_G(V) \subseteq H$. But $F(K) \cap H = 1$ by 12.4(a). Let Z be a minimal W-invariant subgroup of $F(K)$ such that $[V, Z] \neq 1$. Let $Z_1 \subseteq Z$ be an irreducible V-module. Define $R = C_V(Z_1)$. By Clifford Theory, $Z = Z_1 \oplus \ldots \oplus Z_r$ and W permutes transitively the irreducible $V \trianglelefteq W$ modules. Since $R \subseteq Z(W)$, $R \subseteq C_W(Z)$. Thus $Z \in \mathcal{U}(C_G(R))$

If $F(K) = 1$, let E be a minimal W-invariant normal subgroup of K. Then $[V, E] \neq 1$ because otherwise $E \subseteq C_G(V) \subseteq H$ and $E \nsubseteq H$ since $K \in \mathcal{U}(G)$. Let $Y = EV\langle t \rangle$. Then $V \subseteq OF(C_Y(t))$ normalizes every component E_1 of E and induces a cyclic group of automorphisms on each such E_1 by 13.5. Let $R \subseteq C_V(E_1)$, be such that V/R is cyclic. Since $V \subseteq Z(W)$ and W permutes the components of E^W transitively, $R \subseteq C_V(E^W)$. Thus $E^W \in \mathcal{U}(C_G(R))$. Thus (iii) is done.

Let $M = \langle \mathfrak{U}(C_G(R)) \rangle \in \mathfrak{U}(G)$ by (ii). Let $v \in R$ and put $Y_v = \langle \mathfrak{U}(C_G(v)) \rangle$. Then $M \subseteq Y_v \in \mathfrak{U}(G)$.

We show

(iv) $Y = Y_v$ normalizes M.

For since $Y \in \mathfrak{U}(G)$, $F(Y) = Z(Y)$ normalizes $M \subseteq Y$. Let E be any component of $E(Y)$. If $E \subseteq M$, fine. If not, $[V : C_V(E)] \leq p$ as before. If $C_V(E) \supseteq R$, then $E \subseteq E^W \in \langle \mathfrak{U}(C_G(R)) \rangle = M$. Thus $C_V(E) \not\supseteq R$ and so $V = RC_V(E)$. It follows that $[M, C_V(E)] = [M, C_V(E)R] = [M, V]$.

Now $M = F(M)E(M) \in \mathfrak{U}(G)$. As V-group, $F(M) = (C_G(V) \cap F(M))[F(M), V]$. But $C_G(V) \cap F(M) \subseteq H \cap F(M) = 1$ by 12.4(a). Thus $F(M) = [F(M), V]$.

Let L be a component of $E(M)$; V normalizes L as before and $[L, V] \trianglelefteq L$. If $[L, V] \subseteq Z(L)$, then $[L, V] = 1$ as usual. Thus $[L, V] = L$ and $[M, V] = M = [M, C_V(E)]$.

Now $[C_V(E), E, M] = 1$;

$$[E, M, C_V(E)] = 1, \text{ since } E \triangleleft Y = F(Y)E(Y).$$

Therefore $[M, C_V(E), E] = [M, E] = 1$. We have shown that either a component of $E(Y)$ lies in M or it centralizes M. Thus $M \trianglelefteq Y = Y_v$.

(v) $\langle \mathfrak{U}(G) \rangle \in \mathfrak{U}(G)$.

Let $S \in \mathfrak{U}(G)$. First $S = \langle C_S(v) : v \in R^{\#} \rangle$ since R is non-cyclic. Remember $r(V) \geq 3$ and V/R is cyclic and apply 13.5. Now if $v \in R$, $C_S(v) \cap F(S)$ is a W-invariant nilpotent π'-subgroup of S and so

$$C_S(v) \cap F(S) = \langle \mathfrak{U}(C_S(v) \cap F(S)) \rangle.$$

Thus $F(S) = \langle \mathfrak{U}(C_S(v) \cap F(S)) : v \in R^{\#} \rangle$.

On the other hand, if E is a component of $E(S)$, choose $v \in R^{\#}$ such that $C_S(v) \supseteq E$ and then $E^W \subseteq C_S(v)$ and $E^W \subseteq \mathfrak{U}(C_S(v))$. Thus

$$S = F(S)E(S) = \langle \mathfrak{U}(C_S(v)) : v \in R^{\#} \rangle.$$

But then $S \subseteq N_G(M) \subset G$ since $\langle \mathfrak{U}(C_S(v)) \rangle = Y_v \triangleleft M$, for all $v \in V$. Hence $\langle \mathfrak{U}(G) \rangle \subseteq \langle \mathfrak{U}(N_G(M)) \rangle \in \mathfrak{U}(G)$.

Now $W \triangleleft\triangleleft H$. Let $W \triangleleft W_1 \triangleleft \dots \triangleleft W_n = H$. Then W_1 permutes the elements of $\mathcal{U}(G)$ and hence normalizes $\langle \mathcal{U}(G) \rangle = A$, say. Arguing inductively we get H normalizes $A \subseteqq N_G(M) \subset G$ and since H is maximal in G we have first $A \subseteqq H$ and then $A = 1$ since $A \in \mathcal{U}(G)$. This contradiction completes the proof of 14. 7. //

Lemma 14. 8. Let $p \in \pi$, $p \neq 2$, $V \subseteqq O_p(H)$. Let $M \in M(s)$ for some involution $s \in T$, $M \neq H$. Assume $\pi(F^*(H)) \geqq 2$, $V \subseteqq M$ and either

(a) $C_G(V) \subseteqq H$, $V = [V, t]$;

(b) $t \in C_G(v) \subseteqq H$ for all $v \in V^\#$ and V is abelian of type (p, p);

or (c) $t \in C_G(V) \subseteqq H$ and $T \supseteqq W$, where W is of type $(2, 2)$ such that $C_G(w) \subseteqq H$ for all $w \in W^\#$.

Then $O^p(F^*(M)) \subseteqq H$, $O_p(H) \subseteqq M$. If $[s, V] = 1$, then $r(O_p(H)) = 1 < r(O_p(M))$.

Proof. Let $P = O_p(H)$, $Q = O_p(M)$. In case (a), let $V_1 = V^{C_M(t)} \subseteqq O_p(H)$. Thus applying 13. 4 to V_1 we get $[t, V_1] \lhd\lhd F^*(M)$. Since V_1 is a p-group and $V \subseteqq V_1$ in case (a), $V \lhd\lhd F^*(M)$. Thus $V \subseteqq O_p(M) = Q$ and $O^p(F^*(M)) \subseteqq C_G(Q) \subseteqq C_G(V) \subseteqq H$.

In case (b) by 13. 5, $V \subseteqq OF(C_M(t))$ normalizes each component of $E(M)$ and induces a cyclic group of automorphisms in each such component. Thus $O^p(F^*(M)) \subseteqq \langle C_G(v) : v \in V^\# \rangle \subseteqq H$.

In case (c), $V \subseteqq OF(C_M(t))$ normalizes each component of $E(M)$ and centralizes any component of type $L_2(2^n)$ or JR. Thus every component of $E(M)$ of type $L_2(2^n)$ or JR lies in H. Clearly $F(M)_{p'} \subseteqq \langle C_G(w) : w \in W \rangle \subseteqq H$. If a component of $E(M)$ is of type $L_2(q)$, q odd, it is normalized by T and W and by 13. 6 is contained in $\langle C_G(w) : w \in W^\# \rangle \subseteqq H$. Thus $O^p(F^*(M)) \subseteqq H$ in every case.

If $r(Q) \leqq 2$, then since $[P \cap M, O^p(F^*(M))] \subseteqq O^p(F^*(M)) \cap P$, a solvable normal subgroup of $F(M)_{p'}E(M)$. Thus $[P \cap M, E(M)] = 1$ by a familiar argument. Clearly $[P \cap M, F(M)_{p'}] = 1$. Thus $[P \cap M, O^p(F^*(M))] = 1$.

By 12. 4, $P \cap M \subseteqq O_p(M) = Q$. In particular $C_G(C_{Z(Q)}(P \cap M)) = C_G(Z(Q)) \subseteqq M$.

If $r(Q) \geqq 3$, then $|\pi(F^*(M))| \geqq 2$ because otherwise $F^*(M)$ is a p-group and the $F^*(H)$ would be a p-group by 14. 6. This is not the case by hypothesis. If $[s, Q] \neq 1$, then $[s, Q]$ is a $C_G(s)$-invariant subnormal subgroup of $F^*(M)$ and so $[s, Q] \in \mathfrak{A}(M)$ by 14.1(a). Then $C_{Z(Q)}(P \cap M)$ centralizes $[s, Q]$ and is clearly s invariant. By 14.1(b), $C_{Z(Q)}(P \cap M) \in \mathfrak{A}(M)$. Hence $C_G(C_{Z(Q)}(P \cap M)) \subseteq M$.

If $[s, Q] = 1$, then for all $v \in C_{Z(Q)}(P \cap M)$, $C_G(v) \subseteq M$ by 14.7 since $r(Q) \geq 3$.

Thus in every case we have shown that $C_G(C_{Z(Q)}(P \cap M)) \subseteq M$.

Now $V \subseteq P \cap M$ and so $C_G(P \cap M) \subseteq C_G(V) \subseteq H$ and so $C_{Z(Q)}(P \cap M) \subseteq Q \cap H$. Thus $C_G(Q \cap H) \subseteq M$.

It follows that $[O^p(F^*(M)), C_P(Q \cap H)] \subseteq P \cap O^p(F^*(M))$ and, by a now very familiar argument,

$$[O^p(F^*(M)), C_P(Q \cap H)] = 1.$$

We now have $O^p(F^*(M))(Q \cap H)$ acting on P and $O^p(F^*(M))$ centralizes $C_P(Q \cap H)$. Thus $O^p(F^*(M))$ centralizes P. Since $O^p(F^*(M)) \neq 1$, $F \subseteq M$.

We are left only with the last assertion.

Suppose therefore that $[s, V] = 1$. We have already seen that if $r(Q) \leq 2$, then $P \cap M = P \subseteq Q$. Thus if $r(Q) \leq 2$, then $r(Q) \geq r(P)$. Suppose that $r(Q) = r(P)$. Then $\Omega_1(Z(Q))P$ has a subgroup of rank $\geq r(Q)$. It follows that $\Omega_1(Z(Q))P = P$ and so $C_G(P) \subseteq C_G(\Omega_1(Z(Q))) \subseteq M$. Therefore $F^*(H) \subseteq M$ since $P \subseteq M$. Now $O^p(F^*(M)) \subseteq H$ and $P \subseteq Q$. We have $P\,C_Q(P)O^p(F^*(M))(= \bar{S}) \subseteq H$. By 12.4(c), $H = M$, a contradiction. Thus if $r(Q) \leq 2$ then $r(P) < r(Q)$ and so $r(P) = 1$, $r(Q) = 2$.

Assume therefore that $r(Q) \geq 3$ and $[V, s] = 1$. If $[s, Q] \neq 1$, apply 14.5, replacing t by s, H by M, to get a subgroup Q_1 of Q which is $C_G(s)$– invariant such that $[s, Q_1] \neq 1$ and $Q_0 \in \mathfrak{A}(M)$ for all s-invariant subgroups Q_0 of Q_1. Now $V \subseteq C_G(s)$ and so V normalizes Q_1. Let $\bar{V} = [C_{Q_1}(V), s]$. Since V, Q_1 are p-groups, $C_{Q_1}(V) \neq 1$. If $\bar{V} = 1$, then by 2.2, $[s, Q_1] = 1$, a contradiction. Thus $\bar{V} \neq 1$, $\bar{V} = [\bar{V}, s] \subseteq C_G(V) \subseteq H$ and all s-invariant subgroups of Q_1 are elements of $\mathfrak{A}(M)$. Since $|\pi(F^*(M))| \geq 2$, $N_G(\bar{V}) \subseteq M$. We have thus verified that the hypotheses of (a) apply to H, s, $\bar{V}$. It follows that $O^p(F^*(H)) \subseteq M$ and also that $Q \subseteq H$. Thus $F^*(H) \subseteq M$, $F^*(M) \subseteq H$ and $H = M$ by

Theorem A. This is not true and so $[s, Q] = 1$.

Now we find an $\Omega_1(T)P$-invariant normal abelian subgroup $\overline{V}$ of Q of type (p, p). Let A be a maximal abelian normal subgroup of $QP\Omega_1(T)$ contained in Q. If $A \subset C_Q(A)$, we can find $A \subset B \subseteq C_Q(A)$ such that B/A is an irreducible $QP\Omega_1(T)$-module and $B/A \subseteq Z(QP/A)$. Then $\Omega_1(T)$ acts irreducibly on B/A and since T is abelian, $\Omega_1(T)$ of exponent 2, B/A is cyclic. Thus B is abelian, a contradiction, and $A = C_Q(A) \in \mathcal{SCN}(Q)$. If A is cyclic, then Q/A is also cyclic and $r(Q) \leq 2$, not the case. Thus A is non-cyclic and $\Omega_1(A)$ has a chief $QP\Omega_1(T)$ series with cyclic factors. The group $\overline{V}$ has been located successfully.

Any such group $\overline{V}$ lies in an abelian subgroup of type (p, p, p). For since $r(Q) \geq 3$, there exists $X \subseteq Q$ of type (p, p, p). Since $X/C_X(\overline{V}) \subseteq GL(2, p)$, $[X : C_X(\overline{V})] \leq p$. If $C_X(\overline{V}) \neq \overline{V}$, then $\overline{V}C_X(\overline{V})$ is abelian and ≥ 3-generated. If $C_X(\overline{V}) = \overline{V}$, then $\overline{V} \subseteq X$. Apply 14.7 and get $C_G(v) \subseteq M$ for all $v \in V^{\#}$.

If $\overline{V} \subseteq H$, we may apply either conditions (a), (b) to M, $\overline{V}$, s. For if $[s, \overline{V}] \neq 1$, then let $\overline{V}_1 = \langle v_1 \rangle$ where $v_1^s = v_1^{-1}$. Then $C_G(\overline{V}_1) \subseteq M$ and $[\overline{V}_1 s] = \overline{V}_1$. If $[s, \overline{V}] = 1$, then (b) applies directly. Thus $O^p(F^*(H)) \subseteq M$ and $Q \subseteq H$. Hence $F^*(M) \subseteq H$, $F^*(H) \subseteq M$, a contradiction, using Theorem A and $M \neq H$. Hence $\overline{V} \nsubseteq H$.

Now $\overline{V} \trianglelefteq V\overline{V}$ and for some $v \in V$, $C_G(v) \supseteq \overline{V}$. Thus if (b) applies $\overline{V} \subseteq H$, which we have already ruled out. In case (c), $\overline{V} \subseteq \langle C_G(w) : w \in W^{\#} \rangle$ because $W \subseteq \Omega_1(T)$ and again $\overline{V} \subseteq H$. Thus we must be in case (a), $C_G(V) \subseteq H$, $V = [V, t]$.

If $C_P(\overline{V}) \in \mathcal{A}(H)$, then $N_G(C_P(\overline{V})) \subseteq H$ and $\overline{V} \subseteq H$. So $C_P(\overline{V}) \notin \mathcal{A}(H)$.

Now $P' \subseteq C_P(\overline{V})$ and $P' \in \mathcal{A}(H)$ if $P' \neq 1$. Thus if $P' \neq 1$, $\overline{V} \subseteq N_G(P') \subseteq H$, a contradiction. It follows that P is abelian, $[t, P] \neq 1$ because $V \subseteq P$. Apply 14.1(b) and find that every non-trivial t-invariant subgroup of P is an element of $\mathcal{A}(H)$. In particular $C_P(\overline{V}) = 1$ and P acts faithfully on $\overline{V}$. Since $\overline{V}$ is of type (p, p), P is cyclic. //

15. PROOF OF THEOREM A, PART II

Lemma 15.1. $t \in F^*(H)$.

Proof. We show that $[t, O(H)] = 1$ and the result follows from 13.1.

Let $p \in \pi$, $p \neq 2$, $P = O_p(H)$. If $F^*(H) = O_2(H)$, then $O(H) = 1$ and we are done. Assume therefore that $[t, P] \neq 1$.

By 14.5, there exists $\overline{P} \subseteq P$ which is $C_H(t)$-invariant such that $[t, \overline{P}] \neq 1$ and $V \in \mathfrak{A}(H)$ for all t-invariant subgroups V of $\overline{P}$. By 14.4, there exists $R \subseteq T$ such that $r(R) = r(T) - 1$ and $[t, C_{\overline{P}}(R)] \neq 1$. Let $V = [t, C_{\overline{P}}(R)] \in \mathfrak{A}(H)$.

By 14.3, $\{H\} \neq M(s)$ for some involution $s \in R$.

Suppose $M \in M(s)$, $M \neq H$, $s \in R$. If $P \neq F^*(H)$, then 14.8(a) applies to $V \subseteq C_G(R) \subseteq C_G(s) \subseteq M$. Since $[s, V] = 1$, P is cyclic and, since $r(M) > 1$, M is not conjugate to H. Now $[R, V] = 1$ and so R centralizes P, a cyclic group.

Choose $g \in G$ such that $t^g \in R$. Since $[t^g, P] = 1$, $[t, P] \neq 1$, $g \notin H$. Thus $H^g \in M(t^g)$ and $H^g \neq H$. This contradicts the assertion of 14.8.

Thus $P = F^*(H)$ and so by 14.6, $F^*(M)$ is a p-group for all $M \in M(s)$ and for all involutions $s \in G$. In particular $F^*(M)$ is a p-group for $s \in R$, $M \in M(s)$.

By the ZJ-Theorem of Glauberman, $H = N_G(Z(J(S)))$, where S is a Sylow p-subgroup of H. It follows that S is a Sylow p-subgroup of G. Similarly $M = N_G(Z(J(S^g)))$ and so H, M are conjugate.

Now $V \subseteq C_P(R) \subseteq C_G(s) \subseteq M = H^g$.

Choose $U \supseteq V$ maximal such that U is a p-group and $U \subseteq H \cap H^g$, $g \notin H$. Clearly U is not a Sylow p-subgroup of H by the ZJ-Theorem. Thus $N_G(U) \not\subseteq H$.

We break the remainder of the proof into steps.

(i) Every p-subgroup P_1 of G containing a Sylow p-subgroup U_1 of $N_H(U)$ lies in H.

For if P_1 lies in a Sylow p-subgroup S^x of G, then $H \cap H^x \supseteq U_1 \supset U$ and by the choice of U, $x \in H$. Thus $P_1 \subseteq H$.

(ii) $C_G(U) \subseteq U$.

$O_p(N_G(U))U_1 \subseteq H$ by step (i) and if $x \in N_G(U) - H$, then $O_p(N_G(U)) \subseteq H \cap H^x$. Maximality of U forces $U = O_p(N_G(U))$. Now since $V \in \mathcal{A}(H)$, $F^*(C_G(V))$ is a p-group and by 12.6, $F^*(C_G(U))$ is a p-group also. Thus $F^*(N_G(U)) = O_p(N_G(U)) = U$ and so $C_G(U) \subseteq U$. This verifies (ii).

Among all groups $N_G(Y) \supseteq N_G(U)$, where $Y \neq 1$ is a p-group, choose N so that first $|N|_p$ is maximal, and then $|O_p(N)|$ is maximal, and then $|N|$ is maximal. Let $O_p(N) = Z$, $O_{p',p}(N) = X$.

(iii) $X = O_p(N) \times O_{p'}(N)$.

For $O_{p'}(N_X(U)) \subseteq C_X(U) \subseteq U$ by (ii). Hence $N_X(U)$ is a p-group. Since $N_X(U) \trianglelefteq N_G(U)$, $N_X(U) \subseteq O_p(N_G(U)) = U$. Now U normalizes some complement A to $O_{p'}(N)$ in X. Then UA is a p-group. Since $N_G(U) \cap UA = U$, $UA = U$ and $A \subseteq U$. Thus $U \cap X$ is a Sylow p-subgroup of X and so $N_G(U \cap X) \supseteq N_G(U)$ and $|N_G(U \cap X)|_p \geq |N|_p$. Since $U \cap X \supseteq Z$, by the choice of N, $U \cap X = Z$. But $U \cap X$ is a Sylow p-subgroup of X. Hence $X = O_{p'}(N) \times O_p(N)$.

(iv) N is a p-constrained group.

We show that $C_G(Z)$ has odd order and so $C_N(Z)$ is solvable. Hence $C_N(Z) \subseteq O_{p',p}(N)$ as is well known. Let s be an involution in $C_G(Z)$ and take $M \in M(s)$.

If $[s, O_p(M)] = 1$, then $p = 2$, since $O_p(M) = F^*(M)$ and then $F^*(H)$ is a 2-group by 14.6. This is not the case. Thus by 2.2, $[s, C_G(Z) \cap O_p(M)] \neq 1$.

Then $[s, C_G(Z) \cap O_p(M)]$ is a $C_N(s)$-invariant p-subgroup of N. By 13.4, $[s, C_G(Z) \cap O_p(M)] \lhd\lhd F^*(N)$. Hence $[s, C_G(Z) \cap O_p(M)] \subseteq O_p(N) = Z$ and so $[C_G(Z) \cap O_p(M), s, s] = 1$. Apply 0.2 and get a contradiction. Therefore $C_G(Z)$ has odd order and (iv) holds.

We may now apply the ZJ-Theorem to N. Some Sylow p-subgroup S of N contains a Sylow p-subgroup of $N_H(U)$ and so lies in H by step (i). By the ZJ-Theorem, $N = O_{p'}(N) N_N(Z(J(S)))$. But then $Z(J(S)) \subseteq O_{p',p}(N) = O_{p'}(N) \times O_p(N)$ and so $Z(J(S)) \subseteq O_p(N)$. Thus $N = N_N(Z(J(S)))$ and since $|N|$ is maximal, $Z(J(S)) \trianglelefteq N$. But $Z(J(S)) \trianglelefteq H$ since $F^*(H)$ is a p-group. This shows that $N \supseteq H$. But $N \supseteq N_G(U)$ and $N_G(U) \nsubseteq H$. This completes the proof of 15.1. //

Lemma 15.2. $O(F(H))$ <u>has rank at most</u> 2.

Proof. Assume $r(O_p(H)) \geqq 3$ for some odd prime p. Let $P = \Omega_1(Z_2(O_p(H)))$. Since p is odd, P has exponent p. Every normal subgroup of $O_p(H)$ of type (p, p) lies in P and so P is non-cyclic.

If $P/\Phi(P) = P_1/\Phi(P) \times \ldots \times P_k/\Phi(P)$, where $P_i/\Phi(P)$ is an irreducible T-module, define $R \subseteqq C_T(P_1/\Phi(P))$ of rank r(T) - 1. This is possible since T is represented on $P_1/\Phi(P)$ as a cyclic group. If $P_1/\Phi(P)$ is not cyclic, choose $W \subseteqq R$ with r(W) = r(R) - 1. Then clearly $r(C_P(W)) \geqq 2$. If $P_1/\Phi(P)$ is cyclic, then $k \geqq 2$ because $P/\Phi(P)$ and P are not cyclic. Choose

$$W \subseteqq \ker (T \to \text{aut } P_1/\Phi(P)) \cap \ker (T \to \text{aut } P_2/\Phi(P))$$

such that r(W) = r(T) - 2. Again $r(C_P(W)) \geqq 2$ since $r(C_{P/\Phi(P)}(W) \geqq 2$.

Now if $v \in P$, $\langle v \rangle Z(O_p(H)) \trianglelefteq O_p(H)$ since $P \subseteqq Z_2(O_p(H))$. Thus $X = \langle v \rangle \Omega_1(Z(O_p(H)))$ is an elementary abelian normal subgroup of $O_p(H)$. Either $r(X) \geqq 3$, in which case X lies in an element of $\mathcal{SCN}_3(O_p(H))$, or $r(X) \leqq 2$ in which case X still lies in an element of $\mathcal{SCN}_3(O_p(H))$. For by [8] I. 8. 4, $\mathcal{SCN}_3(O_p(H)) \neq \emptyset$, and if $Y \in \mathcal{SCN}_3(O_p(H))$, $|Y/C_Y(X)| \leqq p$. If $C_Y(X) \nsubseteq X$ then $XC_Y(X) \trianglelefteq O_p(H)$ and $r(XC_Y(X)) \geqq 3$. If $C_Y(X) \subseteqq X$, $X \subseteqq Y$ and we are done already.

We can thus apply 14. 7 to find that $C_G(v) \subseteqq H$ for all $v \in P^{\#}$.

By 15. 1, $2 \in \pi(F^*(H))$. Now take $V \subseteqq C_P(W)$ of type (p, p). Since $[t, O_p(H)] = 1$, $t \in C_G(v) \subseteqq H$ for all $v \in V^{\#}$. Also if $s \in W$, then $[s, V] = 1$. Apply 14. 8 and get that if $M \in M(s)$, $M \neq H$, then $r(O_p(H))=1$. This is not true by assumption. Thus $\{H\} = M(s)$ for all $s \in W$.

If W is non-cyclic, there exists a fours-group $W_0 \subseteqq W$ and $t \in C_G(V) \subseteqq H$, where V is a subgroup of $C_P(R)$. Also $s \in R$ and so $[s, V] = 1$. Moreover $C_G(w) \subseteqq H$ for all $w \in W_0$. Apply 14. 8(c) to get that if $M \in M(s)$, $M \neq H$, then $r(O_p(H)) = 1$. This shows that $M(s) = \{H\}$ for all $s \in R$. This contradicts 14. 3.

Thus we have W is cyclic and r(T) = 3. Thus all involutions of G are conjugate.

Consider $W_1 = T \cap F^*(H)$. If V_2 is any subgroup of P of type (p, p), $V_2 \subseteqq C_P(W_1)$ because $[W_1, P] = 1$. Now $t \in C_G(v) \subseteqq H$ for all $v \in V_2^{\#}$ and if $s \in W_1$, $[s, V_2] = 1$. By 14. 8(b), $r(O_p(H)) = 1$, a contradiction unless $\{H\} = M(s)$ for all $s \in W_1$. Of course $t \in W_1$ and so $M(t) = \{H\}$.

If W_1 is non-cyclic, then since $r(T) = 3$, we have an immediate contradiction to 14.3. Thus $W_1 = \langle t \rangle = T \cap F^*(H) \subseteqq Z(H)$ and $H = C_G(t)$.

Now suppose $M \in M(s)$, $s \in R$, $M \neq H$. Then M is conjugate to H because all involutions are conjugate and $M(t) = \{H\}$. So $\{M\} = M(s)$. But $C_P(s) \supseteqq C_P(R) \neq 1$ and so $P \cap M \neq 1$. Thus $C_G(P \cap M) \subseteqq C_G(v) \subseteqq H$, $v \in P \cap M^{\#}$. It follows that $[t, C_G(P \cap M)] = 1$. Now we have $\langle t \rangle \times P \cap M$ acts on $O_p(Z(M))$ and by 2.2, $[t, O_p(M)] = 1$. Thus $O_p(M) \subseteqq H$.

Now choose $V \subseteqq O_p(M)$ of type (p, p). Remember M is conjugate to H and so $r(O_p(M)) \geqq 3$. Then $V \subseteqq H$, $[s, V] = 1$ since $M = C_G(s)$. Interchanging s, t, H, M in 14.8, $s \in C_G(v) \subseteqq M$ for all $v \in V$, V is of type p, p; $[t, V] = 1$. Since $M \neq H$ and $r(O_p(M)) = r(O_p(H)) \geqq 3$, we have a contradiction. //

Lemma 15.3. $H/E(H)$ <u>is solvable.</u>

Proof. We show that $H/C_H(F(H))$ is solvable and so $H/F(H)C_H(F(H))$ is solvable. Since $F(H)C_H(F(H))/F^*(H)$ has odd order, it is solvable. Hence the result.

Let $K = H^{(\infty)}$. Then K acts on $F(H)$ and if $[K, F(H)_p] = 1$, for all p, then $C_H(F(H)) \supseteqq H^{(\infty)}$. Thus choose p such that $[K, F(H)_p] \neq 1$. Since $K = O^p(K)$ there exists a p'-subgroup $X \subseteqq K$ such that $[X, F(H)_p] \neq 1$. Let D be a Thompson critical subgroup of $F(H)_p$ and let $C = \Omega_1(D)$. Of course $p \neq 2$ because $H/C(F(H)_2)$ is odd order and solvable and so $K \subseteqq C_H(F(H)_2)$.

Now by 15.2, $r(O(H)) \leqq 2$ and so $|C| \leqq p^3$. Letting $\overline{C} = C/\Phi(C)$, we have first $C_K(\overline{C})$ is a p-group, and $K/C_K(\overline{C})$ is a subgroup of $GL(2, p)$ and an A^*-group. The only such subgroups are solvable. Thus K is solvable, a contradiction. //

Lemma 15.4. <u>Let</u> K <u>be a component of</u> $E(H)$. <u>If</u> $\{H\} = M(s)$ <u>for every involution</u> $s \in C_T(K)$, <u>then</u> $N_G(T) \subseteqq H$.

Proof. Let $W = C_T(K)$. Assume $N_G(T) \nsubseteq H = M(s)$ for all $s \in \Omega_1(W)$.

(i) T is elementary abelian.

For let $g^{-1} \in N_G(T) - H$. If $W \cap W^{g^{-1}} \neq 1$, then there exists $w, w^g \in W$ involutions and $\{H\} = M(w) = M(w^g)$. It follows that $g \in H$, a contradiction. Since $T = (K \cap T) \oplus W$ and $K \cap T$ is elementary, W is elementary.

Let X, Y be defined as follows:

$$X = N_H(T)/C_H(T), \quad Y = N_G(T)/C_G(T).$$

(ii) $K \trianglelefteq H$.

Since $C_G(T) \subseteq C_G(t) \subseteq H$, $X \subseteq Y$. By assumption $X \neq Y$. If $y \in Y - X$, $W \cap W^y = 1$ and so $|W| \leq [T : W] = |Q|$, where $K \cap T = Q$. In particular there are at most two components of E(H) with Sylow 2-subgroups as large as Q. Moreover if $|T| = |Q|^2$ and E(H) is a central product of two groups isomorphic to K, then H cannot permute the two groups K. It follows that $K \trianglelefteq H$.

Let $|Q| = q$. Choose a subgroup R of K of order $q - 1$ which centralizes W and is regular on Q. We show that

(iii) $X = N_Y(W) \supseteq N_Y(R_0)$ for all subgroups $R_0 \neq 1$ of R.

For if $n \in N_Y(R_0)$, then W^n is centralized by $R_0^n = R_0$. Since R acts regularly on T/W, $W^n = W$ and so $N_Y(R_0) \subseteq N_Y(W)$ because X normalizes K. But if $y \in Y - X$, then $W^y \cap W = 1$. Therefore $X = N_Y(W)$.

(iv) If $\pi(F(Y)) \neq \pi(F(Y) \cap X)$, then $|W| = 2$, $|Q| = 4$.

If $\pi(F(Y)) \neq \pi(F(Y) \cap X)$, let $N \subseteq Z(F(Y))$ be a minimal normal subgroup of Y such that $N \cap X \neq 1$. Then every element of R acts fixed-point freely on N since $C_R(R_0) \subseteq X$ for all subgroups R_0 of R, $R_0 \neq 1$. Put $C = C_T(N)$.

First $C \cap W \subseteq W \cap W^n$, $n \in N$ and so $C \cap W = 1$. Hence $C = [C, R]$ since R is regular on T/W. Because $R \subseteq K$, $C = [C, R] \subseteq K \cap T = Q$. But R is regular on Q and so $C = Q$ or $C = 1$.

Suppose $C = Q$. Then

$$T = [N, T] \oplus C_T(N) = [N, T] \oplus Q.$$

Here $[N, T]$, Q are R-modules and so

$$C_T(R) = (C(R) \cap [N, T]) \oplus C_Q(R).$$

Since $W \cap C = W \cap Q = 1$, it follows that $W \subseteq [N, T]$. But then $W = [N, T]$ and this is impossible because N normalizes $[N, T]$ and $W^n \cap W = 1$ for $n \in N$.

Thus $C = 1$. Now T is a direct sum of faithful and irreducible RN-modules and by [12] 3.4.3, we have $|T| = q|W| = |W|^{q-1}$. This only has solutions if $|W| = 2$, $q = 4$. This verifies (iv).

(v) If W is cyclic, then $O(H) = 1$.

For T is elementary and so $T \subseteq F^*(H)$ since $s \in F^*(H)$ for all $s \in C_T(K)^{\#}$ by assumption and of course $K \cap T \subseteq F^*(H)$. By 12.1, G has a single class of involutions if W is cyclic. Let $s \in T^{\#}$. Since $H = C_G(w)$, $W = \langle w \rangle$, $M = C_G(s)$ is conjugate to H and so $T \subseteq F^*(M)$. Thus $O(M) \subseteq C_G(t)$ because $t \in F^*(M)$ centralizes $F(O(M)) \supseteq C_{O(M)}(F(O(M)))$. Thus $O(M) \subseteq H$. Similarly $O(H) \subseteq M$.

Now K centralizes $O(H)$, a solvable K-invariant subgroup of H and so $F(O(H)) \subseteq OF(C_M(t)) \subseteq O^*(M)$ and $F(O(H)) \subseteq C(T) \cap OF(C_M(t)) \subseteq O(M)$ by 13.5. Thus $F(O(M)) \subseteq O(M) \subseteq H$ and so $F(O(H)) \subseteq F(O(M))$. By symmetry, $F(O(M)) \subseteq F(O(H))$ and $H = M$ if $O(H) \neq 1$. But then $H = C_G(s)$ for all $s \in T$. This is not the case because H is a maximal subgroup of G and $N_G(T) \nsubseteq H$.

(vi) $\pi(F(Y)) = \pi(F(Y) \cap X)$.

If $\pi(F(Y)) \neq \pi(F(Y) \cap X)$, then $|W| = 2$, $|Q| = 4$ by (iv) and $O(H) = 1$ by (v). Thus $F^*(H) = W \times K$ and K is of type $L_2(r)$ where r is odd. Then G is of type JR, a contradiction.

(This is the only time the exact structure of a group of type JR is used.)

(vii) Put $F = R(F(Y) \cap X)$. Then $N_F(R) = F = C_F(R)$.

For $R \trianglelefteq X$, $F(Y) \cap X \trianglelefteq X$. Thus F is a nilpotent group. But R acts regularly on $Q^{\#}$ and so the normalizer of R in the full linear group on Q is, modulo $C_F(Q)$, a subgroup of the multiplicative group of the field $GF(q)$ extended by an automorphism α. If $\alpha \neq 1$, such a group is non-nilpotent and so $F/C_F(Q) = R$. It follows that $F = R \times C_F(Q)$. Hence $R \subseteq Z(F)$.

Let $p \in \pi(F)$, $P = \Omega_1(Z(O_p(F)))$.

(ix) $N_Y(P) \subseteq Z$.

Let $y \in N_Y(P) - X$. Let $x \in C_F(Q)$, $[x, T] \subseteq W$ since $T = W \oplus Q$. Also if $z \in R^{\#}$, $C_T(z) = W$.

First $C_P(Q) \cap C_P(Q)^y = 1$. For if $a, a^{y^{-1}} \in C_P(Q)$, then $[a, T] \subseteq W \cap W^y = 1$ and so $a = 1$.

Also $(P \cap R) \cap (P \cap R)^y = 1$. For if $a, a^{y^{-1}} \in P \cap R$, then $C_T(a) = W = C_T(a^{y^{-1}})$ and $W = W^y$, a contradiction.

Since R is cyclic and P is elementary, $|P \cap R| \leq p$. But $P = P \cap R \oplus C_P(Q)$. It follows that $|P| = p^2$. If S is any subgroup of P such that $P = (P \cap R)S = C_P(Q)S$ then $C_Q(S) \subseteq C_Q(P) = 1$, $C_W(S) \subseteq C_W(P) = 1$.

Thus S acts without fixed points on T. But $P \cap R$, $C_P(Q)$ have fixed points on T and are moved by y. Therefore they must be interchanged. But then y has even order, a contradiction.

We have now shown that $N_Y(P) \subseteq X$ for $P = \Omega_1(Z(O_p(F)))$ for all primes p.

Consider now R_p acting on $F(Y)_p$. If $F(Y)_p \supset (F(Y) \cap X)_p$ choose M such that $(F(Y) \cap X)_p \subset M \subseteq F(Y)_p$ and M is the smallest such R_p-invariant group. Then $[R_p, M] \subseteq (F(Y) \cap X)_p$.

Hence M normalizes $R_p(F(Y) \cap X)_p = F_p$ and so $M \subseteq N_Y(P) = X$. Thus $(F(Y) \cap X)_p = F(Y)_p$. Since $\pi(F(Y) \cap X) = \pi(F(Y))$, we have $F(Y) \cap X = F(Y)$. Now $[R, F(Y)] = [R, F(Y) \cap X] = 1$ and so $R \subseteq F(Y)$. Hence $F = F(Y)$ and $P \trianglelefteq Y$. Thus $X = Y$. This completes the proof. //

Lemma 15.5. Let K be a component of $E(H)$. Then $C_T(K)$ is not cyclic.

Proof. Let $W = C_T(K) = \langle w \rangle$. Since $K \cap T$ is elementary, by transfer $|W| = 2$ and $N_G(T)$ is transitive on involutions of T by 12.1. For $K \cap T$ has a single class of involutions. If $g \in G$ and $t^g \in W$, then $C(w) \subseteq H^g$. But $K \subseteq C_G(w)$ and since $H/E(H)$ is solvable by 15.3,

$K = K^g$. Then $g \in G$ and $t \in W$.

But then $t \in C_T(K) \cap F^*(H)$ by 15.1 and so $t \in Z(H)$. Thus $\{H\} = M(t)$ and so by 15.4, $N_G(T) \subseteq H$. But now transfer gives a contradiction to the simplicity of G. This completes the proof. //

Lemma 15.6. _Assume_ $E(H) \neq 1$. _Let_ K _be a component of_ $E(H)$. _Then_

(a) $C_G(K) \subseteq H$.

(b) _If_ $K \subseteq M \in M(s)$ _for some involution_ $s \in T$, _then_ $M = H$.

(c) $N_G(T) \subseteq H$.

(d) K _is of type_ $L_2(2^n)$.

Proof. If possible choose, H, t, K such that K is of type JR. The proof proceeds by verifying (a), (b), (c) with this restriction on K. When (d) is proved, it follows that this restriction on K is vacuous and the Lemma is completely proved.

Choose an involution $k \in T \cap K$, $M \in M(k)$. By 15.3, $H/E(H)$ and $M/E(M)$ are solvable.

If K is of type JR, let N be the product of all such components of $E(H)$. Otherwise put $N = E(H)$. Note that N is characteristic in any subgroup S of $E(H)$ which contains it.

Let $K_1 \neq K$ be a component of N. Then $K_1 \subseteq C_G(k) \subseteq M$ since $k \in K$. Because $M/E(M)$ is solvable, $K_1 \subseteq E(M)$. Let $E = K_1^{C_M(t)}$. Then $E = E(E)$ lies in $E(M)$ and by 13.3 any component of E, for example K_1, is either a component of $E(M)$ or is of type $L_2(q)$ contained in a component of $E(M)$ of type JR. By choice of H, if K_1 is of type $L_2(q)$, no component of $E(M)$ can be of type JR. Thus K_1 is a component of $E(M)$ and so $K_1 \trianglelefteq E(M)$.

Let $N = KK_1 \ldots K_r$. Then $K_1 \ldots K_r \trianglelefteq E(M) \cap C_G(K) \triangleleft C_G(K)$ $(\subseteq C_G(k) \subseteq M)$. It follows that $N = KK_1 \ldots K_r \trianglelefteq KE(C_G(K)) = E(KC_G(K))$. Hence $E(KC_G(K)) \subseteq N_G(N) = H$ and so $E(KC_G(K)) \subseteq E(H)$. Now N char $E(KC_G(K))$ and so $N_G(E(KC_G(K))) \subseteq N_G(N) = H$. Therefore $C_G(K) \subseteq H$ and (a) holds.

(b) If $K \subseteq M \in M(s)$, then $K \subseteq E(M)$ and as above, K is a component of $E(M)$. Thus by (a) we have $C_G(K) \subseteq M$. Therefore

$E(M) \subseteq H$ and so $E(M) \subseteq E(H)$. But also $E(H) \subseteq M$. Hence $E(H) = E(M)$ and $H = M$.

(c) Let $U = C_T(K)$. By (b), $M(s) = \{H\}$ for all involutions $s \in U$. Then by 15.4, $N_G(T) \subseteq H$ and by 15.5, U is non-cyclic.

(d) Let E be the product of all those components of $E(H)$ not of type $L_2(2^n)$. We show that $E = 1$ and then (a), (b), (c), (d) hold without restriction. Assume $E \neq 1$.

Since U is non-cyclic and $C_G(u) \subseteq H$ for all $u \in U^{\#}$, using $\{H\} = M(u)$ for all $u \in U^{\#}$, every $T \supseteq U$ invariant 2'-subgroup D of G lies in H. Also $D \subseteq E$ for all T-invariant subgroups D of G such that $E(D) = D$ and no component of D is of type $L_2(2^n)$ by 13.6. Remember T normalizes each component of D.

Let $M \in M(s)$, $M \neq H$, $s \in T$. Solvability of $M/E(M)$ implies that $M = O(M)E(M)N_M(T)$. For $TO(M)E(M) \trianglelefteq M$, since M is an A*-group, and then the Frattini argument applies. Since $N_M(T) \subseteq H$, $O(M) \subseteq H$, $E(M) \nsubseteq H$. Let L be a component of $E(M)$ not contained in H. Let $V = C_T(L)$.

By 15.5 applied to M, V is non-cyclic.

If $C_G(v) \subseteq M$ for all $v \in V^{\#}$, then E which is V-invariant, would lie in $E(M)$. Also any component of $E(M)$ not of type $L_2(2^n)$ is $T \supseteq U$ invariant. Thus every component of $E(M)$ not of type $L_2(2^n)$ lies in H. Conversely every component of $E(H)$ not of type $L_2(2^n)$ lies in M. Thus E is the product of all components of $E(H)$ and also $E(M)$ not of type $L_2(2^n)$. Thus $H = M$, a contradiction. Thus there exists $v \in V$ such that $C_G(v) \nsubseteq M$, where v is an involution.

Let $R \in M(v)$. Then $R \neq M$ and of course $L \subseteq R$. Every component of $E(R)$ is T-invariant and every component of $E(R)$ of type JR lies in H by 13.6. Thus every component of $E(R)$ of type JR lies in $E(H)$ and so lies in E. By 14.3 applied to $L^{C_R(s)} \subseteq E(R)$, where clearly $L^{C_R(s)}$ is semi-simple, since $L^{C_R(s)} \subseteq E(M)$, we see that L is a component of $E(R)$ since otherwise L is of type $L_2(q)$ and lies in a component of $E(R)$ of type JR. Since $L \nsubseteq H$, this last possibility does not arise.

But now if K is not of type JR, our restriction on K, t, H is vacuous and so (a), (b), (c) hold all the time. Apply (b) to M, L in place of H, K and get $M = R$, a contradiction. Therefore K is of type JR and so $[T : U] = 8$. But $U \cap V = 1$ for if $x \in U \cap V$, $C_G(x) \subseteqq H$ and $C_G(x) \supseteqq L \nsubseteq H$. Thus $|V| \leqq 8$.

But $|V| = [T : T \cap L]$. Thus $E(M) \supseteqq L$ has at most two components and $E(R) \supseteqq L$ has at most two components. But then $L \trianglelefteq M, R$, a contradiction. //

Lemma 15.7. H __is solvable.__

Proof. If $E(H) \neq 1$, let K be a component of $E(H)$, $U = C_T(K)$. By 15.6, $N_G(T) \subseteqq H$ and $M(s) = \{H\}$ for every involution $s \in U$. Also by 15.5, U is non-cyclic. Hence any U-invariant odd order subgroup of G is contained in H.

Let $M \in M(s)$, $M \neq H$, $s \in T$. Since $M/E(M)$ is a solvable A*-group, $M = O(M)E(M)N_M(T)$. Thus $E(M) \nsubseteq H$.

Let L be a component of $E(M)$ not contained in H and let $V = C_T(L)$. By 15.4, $N_G(T) \subseteqq M$ and $M(v) = \{M\}$ for all $v \in V^{\#}$. Also $U \cap V = 1$ because if x is an involution in $U \cap V$, $L \subseteqq C_G(x) \subseteqq H$.

(i) We may assume that $L \trianglelefteq M$.

If neither $K \trianglelefteq H$ nor $L \trianglelefteq M$, then both $E(H)$ and $E(M)$ contain at least two other components isomorphic to K, L respectively. Then $|U| > |T \cap K| = [T : U]$, $|V| > |T \cap L| = [T : V]$ and then $U \cap V \neq 1$. Thus we have (i).

Since K is of type $L_2(2^n)$, K has a cyclic subgroup R which is inverted by some involution in K and which acts regularly on $T \cap K^{\#}$. Then $U = C_T(R)$ and $R \subseteqq N_G(T) \subseteqq M$. Thus R normalizes L, $L \cap T$ and $C_T(L)$. Moreover R acts irreducibly on $[R, T] = T \cap K$ and so $V \cap [R, T]$ or $V \cap [R, T] = V$.

(ii) $U = T \cap L$ and $RL = R \times L$.

For if $V \cap [R, T] = 1$ then $[V, R] = 1$ because R centralizes T modulo $[R, T]$. But $V \cap C_T(R) = V \cap U = 1$. Thus $V \supseteqq [R, T]$ and if $V \supset [R, T]$, then $C_V(R) \neq 1$ and $C_V(R) \subseteqq U \cap V$.

Thus $V = [R, T]$. It follows that R centralizes $L \cap T$. But $T = V \oplus (L \cap T)$. Since $C_T(R) = U \supseteq L \cap T$ and $U \cap V = 1$, we have $U = L \cap T$.

Moreover R normalizes L, of type $L_2(2^n)$ and centralizes a Sylow 2-subgroup $L \cap T$ of L. Thus R cannot induce field automorphisms on L and must induce inner automorphisms on L. Since $C_L(L \cap T) = L \cap T$, we have $[R, L] = 1$ and $RL = R \times L$.

(iii) U is a Sylow 2-subgroup of $C_G(R)$.

For let $U_1 \supseteq U$ be a Sylow 2-subgroup of $C_G(R)$. Then $U_1 \subseteq C_G(u) \subseteq H$ for $u \in U$ and so $U_1^h \subseteq T \cap C_G(R^h)$ for some $h \in H$. Then $R^h \subseteq K^h$, a component of $E(H)$ of type $L_2(2^n)$ and R^h is a subgroup of order $2^n - 1$ acting regularly on $T \cap K^h$. But $T = (T \cap K^h) \times C_T(K^h)$ and $C_T(K^h) \supseteq C_T(R^h) \supseteq U_1^h$. Since $|C_T(K^h)| = |C_T(K)| = |U|$, we have $|U_1| = |U|$, $U_1 = U$ and U is a Sylow 2-subgroup of $C_G(R)$.

(iv) $U \subseteq F^*(C_H(R))$.

Since $T \cap L$ is elementary by (ii), U is elementary abelian.

By 15.1, $U \subseteq F^*(H)$ because $M(u) = \{H\}$ for all $u \in U$ by 15.6. Thus $U \subseteq C_G(R) \cap F^*(H)$.

Let $F^*(H) = KK_1 \ldots K_rF(H)$;

$$C_G(R) \cap F^*(H) = RK_1 \ldots K_rF(H);$$

$$RF(H) \subseteq F(C_H(R)).$$

Now $K_1 \ldots K_r$ normalizes $F(C_H(R))$, a solvable subgroup of H. Hence $K_1 \ldots K_r$ centralizes $F(C_H(R))$. Since $C_H(R)$ is an A^*-group, it follows that $K_1 \ldots K_r \subseteq F^*(C_H(R))$. Hence $U \subseteq F^*(H) \cap C_G(R) \subseteq F^*(C_H(R))$.

(v) $U \subseteq F^*(C_G(R))$ and so $L \subseteq F^*(C_G(R))$.

For $O(C_G(R))$ is a U-invariant subgroup of G. Thus $O(C_G(R)) \subseteq H$. Hence $[O(C_G(R)), U] \subseteq O(C_G(R)) \cap F^*(C_H(R)) \subseteq F(C_H(R))$. Thus $[O(C_G(R)), U, U] = [O(C_G(R)), U] = 1$. By 13.1, $U \subseteq F^*(C_G(R))$. Since U is a Sylow 2-subgroup of $C_G(R)$ and $L \subseteq C_G(R)$, it follows that $L \subseteq F^*(C_G(R))$. Thus $L = E(C_G(R))$. It follows that $N_G(R) \subseteq N_G(L) = M$. Hence $K = \langle T \cap K, N_K(R) \rangle \subseteq M$. This contradicts 14.6(b) because $M \neq H$. This completes the proof. //

Lemma 15.8. $N_G(T) \subseteq H$.

Proof. Since H is solvable and $t \in F^*(H)$, $t \in O_2(H)$. If $O(H) = 1$, $N_G(T) \subseteq H$ clearly. Let $p \in \pi$, $p \neq 2$. Let P be a maximal T-invariant p-subgroup of G containing $O_p(H)$.

(i) $P \subseteq H$.

For $C_G(P \cap H) \subseteq H$ and so $[t, C_P(P \cap H] \subseteq P \cap O_2(H) = 1$. Since $[P \cap H, t] \subseteq P \cap O_2(H) = 1$, 2.2 implies that $P \subseteq C_G(t) \subseteq H$.

(ii) $\langle \mathcal{N}_G(T, p') \rangle \subseteq H$.

The Transitivity Theorem 4.1 obviously applies here and so $C_G(T) \subseteq H$ acts transitively on the maximal elements of $\mathcal{N}_G(T, p')$. Since $P \subseteq H$, $\langle \mathcal{N}_G(T, p') \rangle \subseteq H$.

Now let $g \in N_G(T)$. Since $H^g \supseteq T$, $F(H^g)_2$ is centralized by T. Because $F(H^g)_{2'} \in \mathcal{N}_G(T, p')$, $F(H^g)_{2'} \subseteq H$ by (ii). Thus $F(H^g) \subseteq H$. It follows that $[t, F(H^g)] \subseteq [t, F(H^g)_{2'}] \subseteq F(H^g)_{2'} \cap O_2(H) = 1$. Because $[t, F(O(H^g))] = 1$, $[t, O(H^g)] = 1$ and so $O(H^g) \subseteq H$. Since H has 2-length 1, being a solvable A*-group, $O(H^g) \subseteq O(H)$. Thus $H^g = H$ and $N_G(T) \subseteq H$. //

Lemma 15.9. $C_G(x) \subseteq H$ <u>for all</u> $x \in O_2(H)^{\#}$. <u>Also</u> $O_2(H)$ <u>is non-cyclic.</u>

Proof. $O_2(H)$ is clearly non-cyclic because, by 15.8, if $O_2(H)$ were cyclic, $O_2(H) \subseteq Z(N_G(T)) \cap T$ and transfer then contradicts the simplicity of G.

Let $x \in O_2(H)$ be an involution, $M \in M(x)$. By 15.8, $N_G(T) \subseteq M$ and so $H = O(H)N_G(T) \subseteq M$. For $[x, O(H)] \subseteq O_2(H) \cap O(H) = 1$ and so $O(H) \subseteq C_G(x) \subseteq M$. //

Lemma 15.10. $M(s) = \{H\}$ <u>for all involutions</u> $s \in T^{\#}$.

Proof. For let $M \in M(s)$, $s \in T$. Then $M = O(M)N_M(T)$. By 15.9, $O(M) \subseteq H$ because $O(M) = \langle C_G(x) \cap O(M) : x \in O_2(H)^{\#} \rangle$. By 15.8 $N_G(T) \subseteq H$. Thus $M = H$. //

This contradicts 13.3 and completes the proof of Theorem A.

APPENDIX: p-CONSTRAINT AND p-STABILITY

These concepts are rather natural generalizations of aspects of the theory of p-solvable groups - see [14]. The definition of p-constraint is taken from a crucial property of p-solvable groups noticed in the famous Lemma 1. 2. 3 of [14]. The definition of p-stability is taken from the famous Theorem B of the same paper. The reader should be familiar with both that paper and also the exposition of these concepts in [12]. A very little discussion of these topics is included to overcome an error in the Gorenstein treatment and also an omission - it is important to know how much induction one has with these concepts.

Definition. Let p be any prime. A group G is said to be p-constrained if, when P is a Sylow p-subgroup of $O_{p',p}(G)$, then $C_G(P) \subseteqq O_{p',p}(G)$.

Definition. Let p be an odd prime, G a group in which $O_p(G) \neq 1$. Then G is said to be p-stable when, for any p-subgroup $A \subseteqq G$ and any A-invariant p-subgroup $P \subseteqq O_{p',p}(G)$ such that $O_{p'}(G)P \trianglelefteq G$ and $[P, A, A] = 1$ it follows that

$$AC_G(P)/C_G(P) \subseteqq O_p(N_G(P)/C_G(P)).$$

First, it is easy to see that if P is a Sylow p-subgroup of $O_{p',p}(G)$ and $C_G(P)$ is p-solvable, then G is p-constrained. For by Lemma 0. 3, $C_{G/O_{p'}(G)}(P) = C_G(P)O_{p'}(G)/O_{p'}(G)$ and there is no loss of generality in assuming $O_{p'}(G) = 1$. Thus $C_G(P) \trianglelefteq G$ and we can find $K \trianglelefteq G$, $K \supset P$ such that $K \subseteqq PC_G(P)$ and K/P is a chief factor of G/P, if $C_G(P) \nsubseteq P$. Since $P = O_{p',p}(G)$, K/P is a p'-group and then a q-group for some prime $q \neq p$. Since $K \subseteqq PC_G(P)$, $K = PC_K(P)$. Let Q be a Sylow q-subgroup of K contained in $C_K(P)$. Then $K = PQ = P \times Q$ and so $Q \subseteqq O_{p'}(K) \subseteqq O_{p'}(G) = 1$. This shows that G is p-constrained if $C_G(P)$ is p-solvable where $O_{p'}(G)P = O_{p',p}(G)$, P a p-group.

Notice however that the property of p-constraint does not necessarily pass to either subgroups or to factor groups. In order to see this consider $A_5 \times C_3 \subseteqq A_8$. Clearly $A_5 \times C_3$ is not 3-constrained. We can make a group $7^8(A_5 \times C_3)$ by extending an elementary abelian group of order 7^8 by $A_5 \times C_3$ with the natural action. Now let this new group act faithfully on any elementary abelian 3-group. The group $G = 3^k 7^8(A_5 \times C_3)$ so constructed is clearly 3-constrained since the elementary abelian group $3^k = O_{3', 3}(G)$ is self centralizing. Since $A_5 \times C_3$ is both a subgroup and a factor group of G, we see that p-constraint does not induct in either direction.

The following result shows that some induction to both factor groups and subgroups is possible under certain circumstances.

Lemma 1. (a) A group G is p-constrained if and only if $G/O_{p'}(G)$ is p-constrained.

(b) If G is p-constrained, then $N_G(P)$ and $C_G(P)$ are p-constrained for every p-subgroup P of G.

Proof. (a) This follows immediately from 0.3.

(b) Using 0.3 and (a) we may assume that $O_{p'}(G) = 1$. The result follows from 12.5.

We turn now to p-stability. It is worth stating the celebrated Theorem B here so that the original genesis of p-stability can be discussed.

Theorem B (P. Hall and G. Higman). Let G be a p-solvable group of linear transformations in which $O_p(G) = 1$, acting on a vector space V over a field F of characteristic p. Let x be an element of order p^n. Then the minimal polynomial of x on V is $(X - 1)^r$ where

(i) $r = p^n$ or

(ii) there exists an integer $n_0 \leqq n$ such that $p^{n_0} - 1$ is a power of a prime q for which the Sylow q-subgroups of G are non-abelian. In this case, if n_0 is the least such integer, then

$$p^{n-n_0}(p^{n_0} - 1) \leqq r \leqq p^n.$$

Of course, it is no surprise that the minimal polynomial is of the form $(X - 1)^r$ where $r \leqq p^n$. Clearly the minimum polynomial divides

$X^{p^n} - 1 = (X - 1)^{p^n}$. The interesting part of this Theorem is the lower bound. For our purposes we will enquire when the constant r can be 2, i.e., when will x have quadratic minimum polynomial? Clearly it always will if $p^n = 2$ and this is the reason for excluding the prime $p = 2$ from the definition of p-stability.

Thus $r = 2$ occurs only when $p^{n-n_0}(p^{n_0} - 1) = 2$ and this holds only when $n = n_0 = 1$, $p = 3$. A Sylow 2-subgroup of G will be non-abelian and $3/|G|$. A careful reading of the proof of Theorem B shows that $r > 2$ unless $SL(2, 3)$ is involved in G.

To get to the hypothesis of Theorem B in an abstract group G, suppose that P is a p-subgroup of G and that $O_p(N_G(P)/C_G(P))=1$. Let A be an abelian p-subgroup of $N_G(P)$ and let $V = P/\Phi(P)$. Then $[P, A, A] = 1$. If $v \in V$, $A \in A$, $[v, a] = -v + v^a = v^{-1+a}$,

$$[v, a, a] = v^{(-1+a)^2}.$$

Since $[P, A, A] = 1$, $(-1 + a)^2$ is the zero endomorphism of V and so a acts on V with at worst quadratic minimum polynomial. If for example $SL(2, 3)$ is not involved in G, we know that a must act trivially on V. It will follow that $A \subseteqq P$ and so P contains every abelian subgroup normal in a Sylow p-subgroup of $N(P)$. Further information can be found in [12].

Once again, p-stability does not go over to proper sections.

For a careful reading of the proof of 3.8.3 of Gorenstein [12] shows that, when x and y are conjugate p-elements in a group X such that $\langle x, y\rangle$ is not a p-subgroup while x acts on a G-vector space of characteristic p with quadratic minimum polynomial, it follows that $\langle x, y\rangle$ involves $SL(2, p)$.

Now consider, A_8, the alternating group of degree 8. Let V be an elementary abelian 3-group of order 3^8 and G the split extension of V by A_8 with the natural action of A_8 on V. Given any 3-element x of A_8 there is always a conjugate y of x in A_8 such that $\langle x, y\rangle \cong A_4$. Hence $\langle x, y\rangle$ does not involve $SL(2, 3)$ and so x cannot act on V with quadratic minimum polynomial. Thus G is 3-stable.

Since however G contains a section isomorphic to $Qd(3)$, a split extension of a group of type (3, 3) by SL(2, 3) and a classical non-3-stable group, we have exhibited a 3-stable group with a non-3-stable proper section.

In view of this, the following Lemma is useful.

Lemma 2. _A group_ G _in which_ $O_p(G) \neq 1$ _is p-stable if and only if_ $G/O_{p'}(G)$ _is p-stable._

Proof. If G is p-stable, write $\overline{G} = G/O_{p'}(G)$, etc. Suppose that $\overline{A} \subseteq N_{\overline{G}}(\overline{P}) = \overline{G}$ is such that $[\overline{P}, \overline{A}, \overline{A}] = 1$ where $\overline{A}$, $\overline{P}$ are p-subgroups of $\overline{G}$. Let $K = O_{p'}(G)$. Now A normalizes PK and since the number of Sylow p-subgroups of PK is prime to p, there exists a $k \in K$ such that A^k normalizes P.

Also $[P, A^k, A^k] \subseteq K$. Hence

$$[P, A^k, A^k] \subseteq P \cap K = 1.$$

Therefore since G is p-stable, $A^k C_G(P)/C_G(P) \subseteq O_p(N_G(P)/C_G(P))$. By 0.3, $C_{\overline{G}}(\overline{P}) = C_G(P)K/K$, and so

$$\begin{aligned}\overline{G}/C_{\overline{G}}(\overline{P}) &\cong G/C_G(P)K = N_G(P)C_G(P)K/C_G(P) \\ &\cong N_G(P)/(N_G(P) \cap C_G(P)K) = N_G(P)/C_G(P).\end{aligned}$$

Since $A^k C_G(P)/C_G(P) \subseteq O_p(N_G(P)/C_G(P)) \cong O_p(\overline{G}/C_{\overline{G}}(\overline{P}))$ and $\overline{A^k} = \overline{A}$, $C_{\overline{G}}(\overline{P}) = \overline{C_G(P)}$, we have that $\overline{G}$ is p-stable.

Conversely suppose that $\overline{G} = G/O_{p'}(G)$ is p-stable. Let P be a p-subgroup of G such that $O_{p'}(G)P \subseteq G$ and suppose that $A \subseteq N_G(P)$ is a p-subgroup such that $[P, A, A] = 1$. Since $[\overline{P}, \overline{A}, \overline{A}] = 1$, we have that

$$\overline{A}C_{\overline{G}}(\overline{P})/C_{\overline{G}}(\overline{P}) \subseteq O_p(\overline{G}/C_{\overline{G}}(\overline{P})).$$

Now

$$C_{\overline{G}}(\overline{P}) = C_G(P)O_{p'}(G)/O_{p'}(G)$$

and

$$N_G(P).C_G(P).O_{p'}(G)/C_G(P).O_{p'}(G) = G/C_G(P)O_{p'}(G)$$

$$\cong N_G(P)/(N_G(P) \cap (C_G(P)O_{p'}(G)) = N_G(P)/C_G(P).$$

Thus under the above isomorphism

$$AC_G(P)/C_G(P) \subseteqq O_p(N_G(P)/C_G(P))$$

and G is p-stable. Lemma 2 is completely proved.

References

[1] J. Alperin and R. Lyons. On conjugacy classes of p-elements, J. Algebra 19 (1971), 536-7.

[2] H. Bender. Über den grössten p'-Normelteiler in p-anflösbaren Gruppen. Arch. Math. (Basel) 18 (1967), 15-16.

[3] H. Bender. On the uniqueness theorem. Ill. J. Math. 14 (1970), 376-84.

[4] H. Bender. Transitive Gruppen gerader Ordnung, in denen jede Involutionen genau einen Punkt festlässt. J. Algebra 17 (1971), 527-54.

[5] H. Bender. On groups with abelian Sylow 2-subgroups. Math. Zeitschr. 117 (1970), 164-76.

[6] N. Blackburn. Automorphisms of finite p-groups. J. Algebra 3 (1966), 28-29.

[7] R. Brauer. On groups of even order with an abelian Sylow 2-subgroup. Arch. Math. 13 (1962), 55-60.

[8] W. Feit and J. G. Thompson. Solvability of groups of odd order. Pacific J. Math. 13 (1963), 755-1029.

[9] G. Glauberman. A characteristic subgroup of a p-stable group. Canad. J. Math. 20 (1968), 1101-35.

[10] D. M. Goldschmidt. A group theoretic proof of the $p^a q^b$-theorem for odd primes. Math. Zeitschr. 113 (1970), 373-5.

[11] D. M. Goldschmidt. Strongly closed abelian 2-subgroups of finite groups. (Preliminary announcement, Proc. Gainesville Group Theory Conf. North Holland Math. Studies 7, 1972.) Appeared as 2-Fusion in finite groups. Annals of Math. 99 (1974), 70-117.

[12] D. Gorenstein. Finite groups. Harper and Row, New York 1968.

[13] D. Gorenstein and J. H. Walter. The characterization of finite groups with dihedral Sylow 2-subgroups. J. Algebra 2 (1965),

85-151, 218-70, 354-93.

[14] P. Hall and G. Higman. On the p-length of a p-solvable group and reduction theorems for Burnside's problem. Proc. Lond. Math. Soc. 6 (1956), 1-40.

[15] B. Huppert. Endliche Gruppen I. Springer Verlag, Berlin, 1967.

[16] Z. Janko. A new finite simple group with abelian Sylow 2-subgroups and its characterization. J. Algebra 3 (1966), 147-86.

[17] Z. Janko and J. G. Thompson. On a class of finite simple groups of Ree. J. Algebra 4 (1966), 274-92.

[18] H. Matsuyama. Solvability of groups of order $2^a p^b$. Osaka J. Math. 10 (1973), 375-8.

[19] R. Ree. A family of simple groups associated with a simple Lie algebra of type G_2. Am. J. Math. 83 (1961), 401-20.

[20] M. Suzuki. A new type of simple groups of finite order. Proc. Nat. Acad. Sci. US. 46 (1960), 868-70.

[21] J. G. Thompson. Towards a characterization of $E_2^*(q)$. J. Algebra 7 (1967), 406-14.

[22] J. H. Walter. Finite groups with abelian Sylow 2-subgroups of order 8. Inventiones Math. 2 (1967), 332-76.

[23] J. H. Walter. The characterization of finite groups with abelian Sylow 2-subgroups. Ann. Math. 89 (1969), 405-514.

[24] H. N. Ward. On Ree's series of simple groups. Proc. Am. Math. Soc. 6 (1963), 534-41.